T0073733

CONTENTS

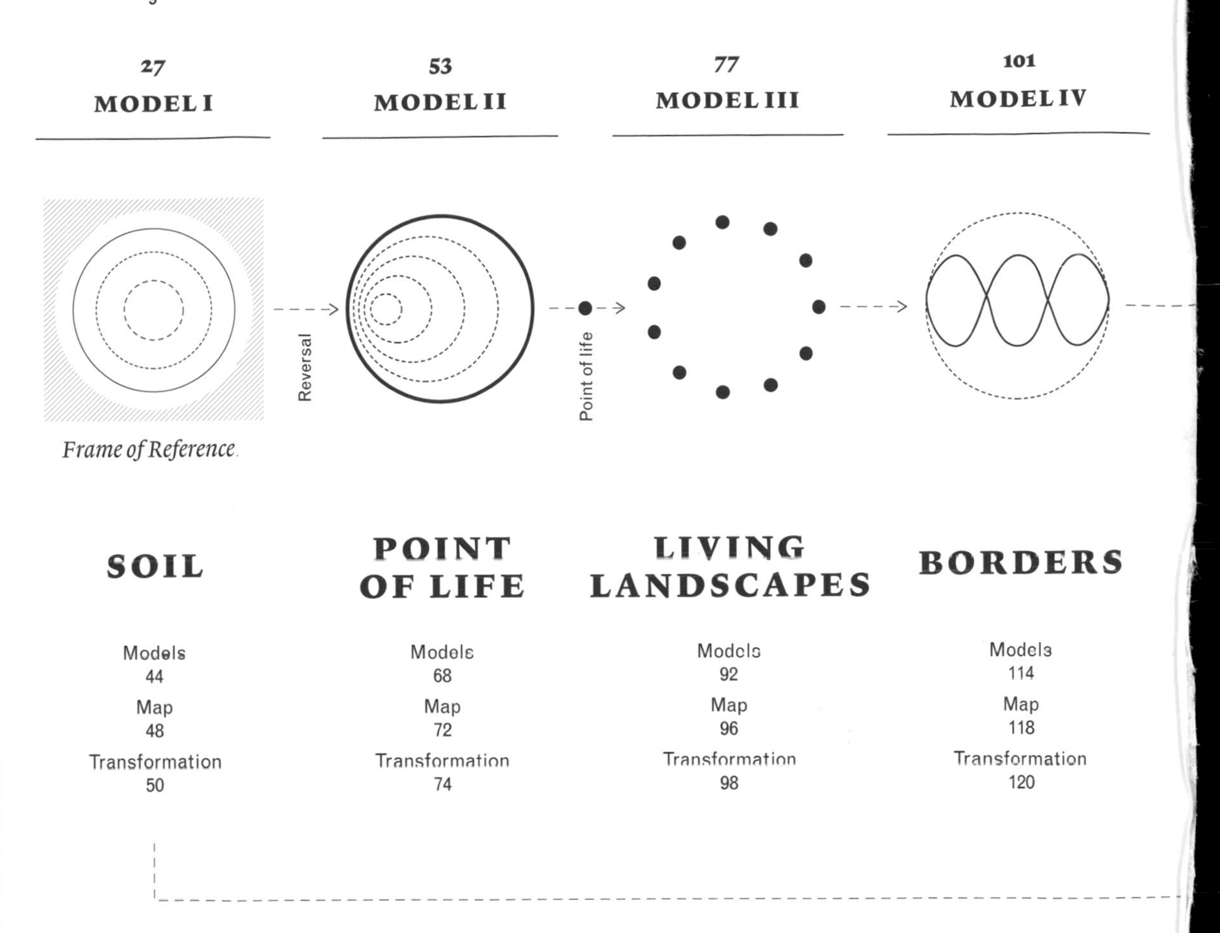

Frame of Reference

Frédérique Aït-Touati,
Alexandra Arènes,
and Axelle Grégoire

TERRA FORMA

A Book of
Speculative Maps

foreword by Bruno Latour
translated by Amanda DeMarco

The MIT Press
Cambridge, Massachusetts
London, England

TERRA FORMA

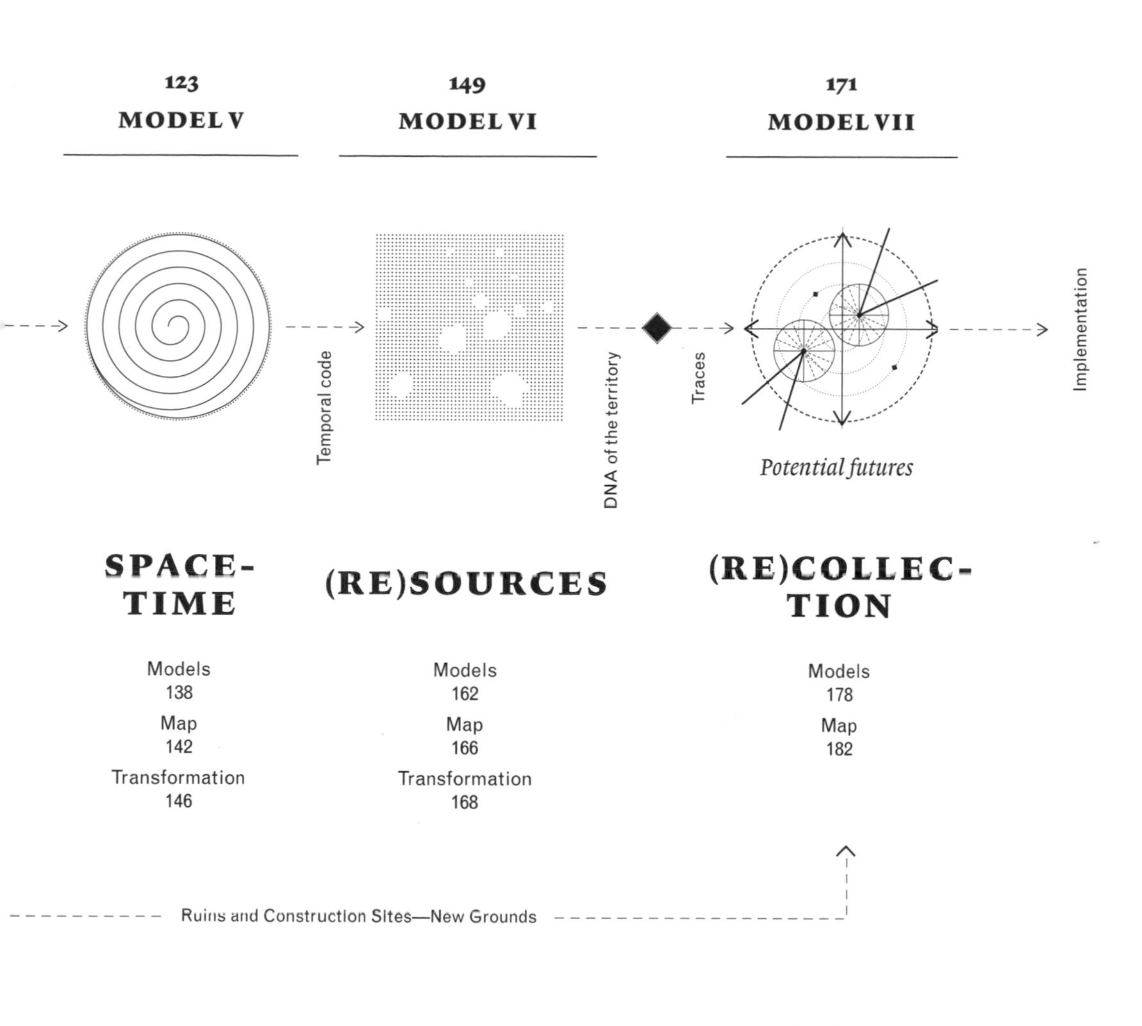

Temporal code
DNA of the territory
Traces
Potential futures
Implementation

FOREWORD: HOW TO FIGHT ABOUT SPACE

This is a strange moment when people are beginning to wonder where, when, and who they are. There is a new uncertainty about the shape of the land, about this moment in history, about the role we humans should play. No wonder we suddenly need tools to orient ourselves, to envision how to cope and where to settle. At a time of ecological mutations, a new map of the Earth seems indispensable. But what sort of map? That's the question this remarkable UFO of a book is tackling head on.

The word "map" quickly brings to mind the idea of a grid that defines the *base map* (what in French we call *le fond de carte*) for everything that will be added later, layer upon layer. But "map" originally described something resembling nothing more grandiose than a marked-up paper towel or napkin. When you scribble doodles on a napkin, you have produced a map—according to the word's Latin etymology. And you have provided a pretty good example of its most useful function: in the course of a conversation, someone grabs a flat piece of paper and draws symbols and features that make sense to those *engaged* in the discussion, even though those scribbles, figures, and notations might not be at all legible to an outsider.

Just who are those *outsiders* who claim to be able to read all maps as if they were designed for them? Why should they enjoy an undisputed hegemony regarding the definition of the activity of mapping?

In a beautiful show a few years ago titled "When Artists Drew Maps," French archivists presented to the public a beautifully painted set of maps—flat pieces of parchment—that utilized none of the conventions that were imposed in the seventeenth century to define the grid-limited base map.[1] Far from being meant for outsiders, those marvelous drawings, painted by artists and not by surveyors, were intended for a very particular audience, and they were checked and certified by the parties in conflict who had commissioned them in order to enlighten arbiters as to the disputes. Those documents, sometimes able to superimpose conflicting views of the same piece of land, would be signed by both parties after protracted and contradictory visits to the site; they were magnificently called *figures accordées*, "agreed-upon figurations."

A few decades later, artists were kicked out of the trade and replaced by surveyors and geometers under the firm hand of the monarchy. At that point maps were devised to help complete outsiders ease their way through unknown places they might wish to dominate and control without having to agree, discuss, or negotiate in any way with the locals. Maps and the colonial imaginary were now well and jointly ensured.[2]

The maps in the present book, drawn by architects and urban planners, are much closer to "figures accordées" than to grid-paralyzed base maps. There are no across-the-board conventions that can be transported from place to place, time to time, or scale to scale. One cannot extract those drawings from the cases at hand. They are not made to help arbiters settle a dispute. They are made instead to direct attention to new features of the land by enriching, chapter after chapter, the meaning of "space." In the hands of these three authors, space is the least obvious feature of figurations; but it is the one most in need of the invention of some ad hoc grammar. Cartesian coordinates,

---

1 J. Dumasy, N. Gastaldi, and C. Serchuk, *Quand les artistes dessinaient les cartes: Vues et figures de l'espace français, Moyen Âge et Renaissance* (Paris: Archives Nationales et Éditions le Passage, 2019).
2 Something the remarkable site decolonial atlas, https://decolonialatlas.wordpress.com, tries every day to reverse, tweet by tweet.

conventionally understood, could in no way be chosen as the ultimate repository.

"Space" is a word just as misleading as the word "map," especially when it appears in the widely used expression "spatial turn." There is no question that this book is about a "spatial turn" too, except that what is meant by space is in direct opposition to the usual notion of spatial relations illustrated by the gridded base map.

Cartesian coordinates are ideal ways to register data sets. So what geographers mean by taking the spatial turn is gathering as much data as possible to show the relative *distance* between data points. The result is the homogenization of all types of relations under the metric of relative distance. Adding to this oversimplification, the shift from the orthographic to the digital (to use John May's vocabulary) throws one as far as possible from anything "spatial," if we mean by this the intertwined forces that shape the Earth.[3]

When geographers, rather late in the twentieth century, enthusiastically embraced the spatial turn, or rather the easy access to GIS data structures, they *lost space* and were largely lost "in space."[4] The hard, complicated, boots-on-the-ground, contradictory, specific, tailor-made attention to the "geo-" that the suffix "-graphy" underlined was jettisoned in the name of a more "scientific," data-driven management to help outsiders drive through a land in which they had no real interest—except for locating resources to be exploited. Geographers lost the Earth in the process. This book is an attempt to retrieve it by following other peculiarities of the "spatial": the fact that it is

3 John May, *Signal. Image. Architecture* (New York: Columbia Books on Architecture and the City, 2020).
4 Olivier Orain, "La géographie comme science: Quand 'faire école' cède le pas au pluralisme," in Couvrir le monde: Un grand XXe siècle de géographie française, ed. M. C. Robic (Paris: ADPF/La documentation française, 2006), 90–123.

based on overlapping, superposed, multidimensional, time-dependent entities. So, paradoxically, each chapter of this book does indeed advocate taking a spatial turn by revolting against another, earlier version of that turn.

It is not surprising that the draftswomen of this book are accompanied in their quest by a historian of science whose specialty is to understand how much of what is often called "the scientific revolution" depended on *fiction* only to deny any connection with that fiction later on. Having studied a canonical case of a first spatial turn—how the planet Earth moves through space—is a good way to tackle the new question of how the Earth, as a Critical Zone, is turning against itself.[5] Understanding how the Earth as a planet was supposed to behave in the infinite universe required an immense effort of imagination until the "modern world" was placed "in" the space of grid-based maps. It does not require great critical skills to understand that the task today is just as immense—except now our task is to reverse those old habits of thought and to resume the exploration of an Earth that no longer resides "in the same space." It has become tricky to situate "where" something is from the outside when overlapping spatial relations are actually not the frame but the *results* of how living organisms interact and fight with one another.

One of the strangest effects of this set of tentative explorations is to invert the various meanings of "spatial." In the chapter called "Resources," we can still find what would be called a normal definition of spacc—except it is no longer what makes up the base map, but is instead the result of a highly specific and fully localized set of practices the authors call *suction cups*. What a great phrase! And an excellent political move as well. To be sure, in some places a now traditional (meaning

5 Two of the authors also contributed to a book that was very much inspired by their work: Bruno Latour and Peter Weibel, *Critical Zones: The Science and Politics of Landing on Earth* (Cambridge, MA: MIT Press, 2020).

modern) definition of space is still being implemented, but they are contained and thereby *limited* loci. And they are just those places where the conflicts are the deepest between the extractivists who activate the isolating and strictly controlled suction cups and those who wish instead to reveal the way the Earth *forms itself*. *Terra Forma* is as far as possible from the conventional understanding of "terraforming," the ideal of some investors that describes what must happen within the confines of those suction cups before they are deployed on the Moon or on Mars when there is nothing left to extract from the Earth. *Terra Forma* provides another—particularly creative—meaning of the expression "the shape of the land that shapes itself." This book is another instance of the new politics of space: how you understand "where" you are will define what sort of politics you are going to sustain.

Bruno Latour

LIVING AMONG THE LIVING

This book tells the story of the exploration of an unknown world—our own. Following in the footsteps of the Renaissance travelers who set out to map the *terra incognita* of the New World, we are endeavoring five centuries later to discover a new Earth, or rather to rediscover, in a new way, the Earth we think we know so well. But we no longer live in the Age of Discovery. Our voyage leads inward and not into the distance, into thickness rather than across expanses. Whereas the task of cosmographers was to expand the horizon, making cartography an art that united movement and traces, our mission is in some sense the opposite: we've changed course, moving from the horizon line to the thickness of the ground, from the global to the local. We've also changed our pace, posture, and tone. Necessity and urgency have replaced the scientific passions of curiosity and discovery. The feeling of having an unlimited world to conquer—the *plus ultra* of Charles V, explorers, and Francis Bacon—has been replaced by a growing awareness of "planetary boundaries." The innocence of the first travelers has been lost; we now know that such expeditions are fatal. We know about the conquests and land grabs that they led to. But our aim is not to wallow in the ruins, nor to abandon the task of discovery. According to the researchers—ethologists and ethnologists, geochemists and biologists—who constantly repopulate our world, shedding light on its new dimensions, it seems that we are much more numerous and much more diverse than we had thought, and that the limits of the world are not what we had imagined them to be. Take, for example, the ground on which we live without knowing what it is made of or what populates it. For decades, the actions of animate entities, rocks, and landscapes have called into question our old ways of thinking about and acting on the land; we can no longer ignore the Earth's response to our own activities, which manifests with increasing vehemence and speed.

How can we inhabit this world made up of lives other than ours, this reactive Earth? Maps as we know them bespeak a relationship to a space emptied of life, an available space that can be conquered or colonized. We had to begin by trying to repopulate maps. To do so, we have shifted the object of notation, trying to delineate not the soil without living things, but the living things in the ground, the living of the soil, as they constitute it. This cartography of the living attempts to document the living as well as their traces, to generate maps based on bodies, rather than on topography, frontiers, and territorial borders (fig. 1).

In the past few years, a new type of map has appeared that seems to react to the activity of actors: GPS maps.[1] Our notion of space and of movement has been profoundly influenced by this tool, which locates the positions of a point on a fixed background map by using data from multiple satellites in orbit. GPS tracks the activities of actors, captures the movements of living things on a predefined map, and allows them to orient themselves in space. Everyone can generate their own route, their own path,

---

1 Along with GPS (Global Positioning System), new types of mapping tools include Google Maps and Google Earth (democratized tools of geolocalization but also route planning), and the French government initiative Géoportail (an institutional and administrative open access tool that allows users to obtain land registers and census information from historical maps), which also takes part in the open data movement. The geographic information system used by state institutions (communities and others) should also be mentioned, a professional tool that claims to be open source (QGIS, freeware), but whose data are often secured and complicated to manipulate. It is a system for accumulating data based on geographical coordinates—a true encoding of territories. In an era of the democratization of maps and the development of open data, the problem is no longer access to information but rather its organization. The open source movement has revolutionized our use of maps as well as our professional practices. Architects, city planners, and landscapers can no longer do their work without Google Maps.

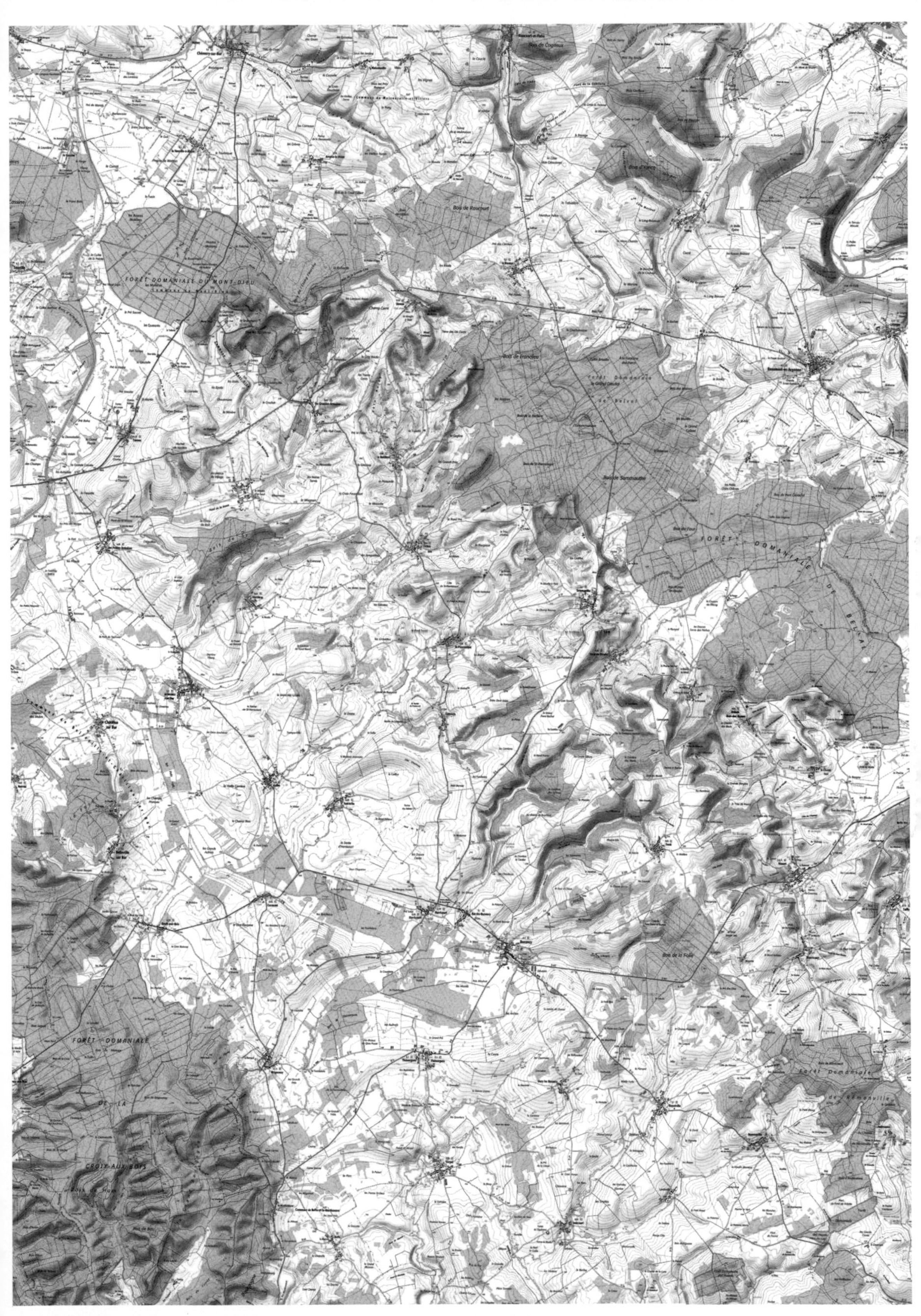

fig. 1 Example of a map from the French National Geographical Institute

navigating within a space whose parameters are determined *a priori*. These maps no longer have much in common with earlier maps, which had to be drawn by walking the terrain oneself, or even by hearing about it from others returning from a journey. Satellites offer precision geodesy, and they help to contour lines, signal evolutions, and reveal the metamorphosis of territories. But they lack narrative, an assemblage of stories told, a multiplicity of people and narrators that allow a map to become a synthesis, to be simultaneously unique and multiple (fig. 2).

Would it be possible to preserve this wonderful transformation in our conception of maps—focusing on the living and not on a fixed location—and also to invent a tool that would allow us not only to follow the trajectories of living things, but also to understand how they shape spaces and even constantly produce them? Living things do much more than move around; they manipulate air and matter in order to create the conditions for their survival, sometimes in cooperation with other living things that need similar or complementary conditions, sometimes in conflict with other collectives, or even in a counterproductive or self-destructive way, as in the case of humanity and the pollution it generates. In this new kind of GPS, the living things themselves create the space, defining their own parameters and producing the map (fig. 3). This changes the status of the map; it is no longer a fixed image, but rather a provisional state of the world, a working tool in evolution, constantly generated by living things. It also changes the status of space: it is no longer a simple container, but rather a living, resonant milieu, composed of thousands of superimpositions and actions of the beings that surround us, constantly and indefinitely produced by the movements and perceptions of those who create it.[2]

---

2 The relationship between the map and the territory it depicts is a classic question: is a map a representation of the world, or rather, as John Pickles suggests, an inscription that functions (or doesn't) in the world and modifies it? Do maps

This experimental undertaking is the work of six hands, a collaboration between two architects dedicated to landscape and strategic territorial design (Alexandra Arènes and Axelle Grégoire) and a historian of science and theater director (Frédérique Aït-Touati). United by our fascination with the way maps can unfold entire worlds, together we decided to seize upon the urgent crisis of how to represent a world in the throes of disruption. To do so, we had to invent shared methods and a shared language. Our book is the result of a series of intense discussions, manipulations of concepts, exchanges of images and ideas, relay handovers, translations, and rebound effects. There are so many hybridizations that it is now difficult to determine the origin of a particular paragraph, a particular line in a drawing, or a particular idea. As much about the exploration of familiar territories as it is about our methods, this experiment is characterized by reworking tools (maps, cross sections, exploration tales), borrowing from other disciplines, and reinterpretation. The objective is to develop new methods inspired by modeling, the preferred tool of contemporary science for anticipating possible futures (climate models), as well as for formalizing abstract ideas (mathematical models).

Each tool, or model, is conceived as a new focal point through which a territory is redrawn. We sought tools that would allow us to navigate between scales, to escape the established

---

precede the territory that they represent, and can they be understood to produce it? Many studies have suggested completely rethinking our habitual notions about the emergence of territories, as well as our traditional understanding of the role of maps. Various schools focus, for example, on the political and institutional implications of cartography (John Pickles, *A History of Spaces: Cartographic Reason, Mapping, and the Geo-Coded World* [London: Routledge, 2004]), on various practices that lead to the production of maps (Jeremy Crampton, "Cartography: Performative, Participatory, Political," *Progress in Human Geography* 33, no. 6 [May 21, 2009]; Martin Dodge et al., *Rethinking Maps: New Frontiers in Cartographic Theory* [London: Routledge, 2009]), and on the performative uses of maps (Denis Cosgrove, ed., *Mappings* [London: Reaktion Books, 1999]). We inscribe our work in this vast current of reflections on maps and their uses.

1 The maps are drawn based on tales of exploration. Explorers travel the world and return to tell of their journey: the lands discovered, the people encountered there, the available resources . . .

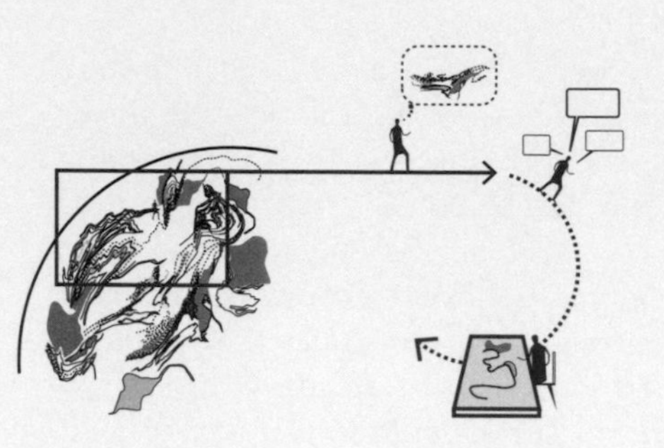

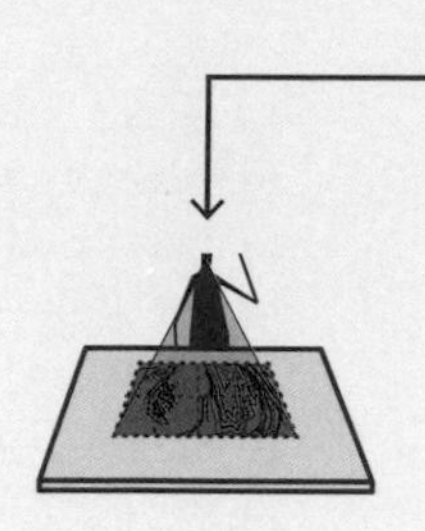

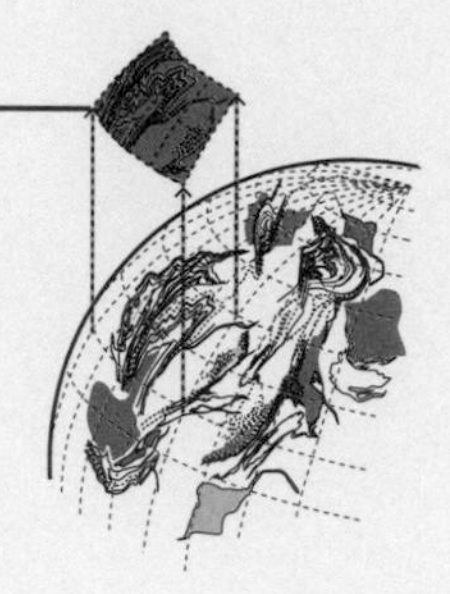

2 The cartographer drafts the maps in their workshop and then distributes them to the next explorers.

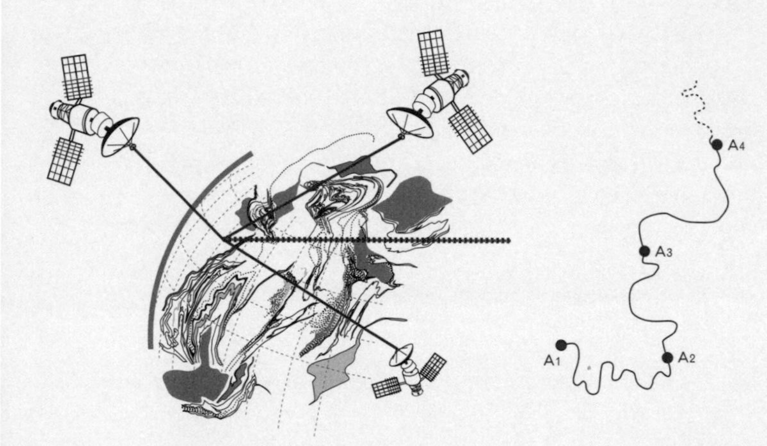

3 Technological progress: satellites around the planet *map* the Earth. Aerial photographs are sent back to Earth and reconstituted on giant IT servers. Geolocalization is invented: any device with a signal traceable by satellite can be located precisely by its latitudinal and longitudinal coordinates.

4 Satellites produce maps. The cartographer's drawing is replaced by the machine's "objective" bird's-eye view. Maps are now accessible everywhere, by anyone, and can be used for many tasks, some personal (orienting oneself) and some professional (analyzing and planning the future of territories).

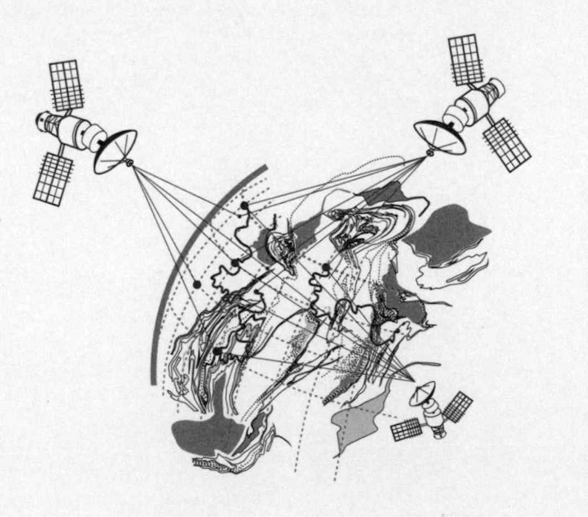

fig. 2 Brief history of mapmaking and introduction of the Terra Forma method

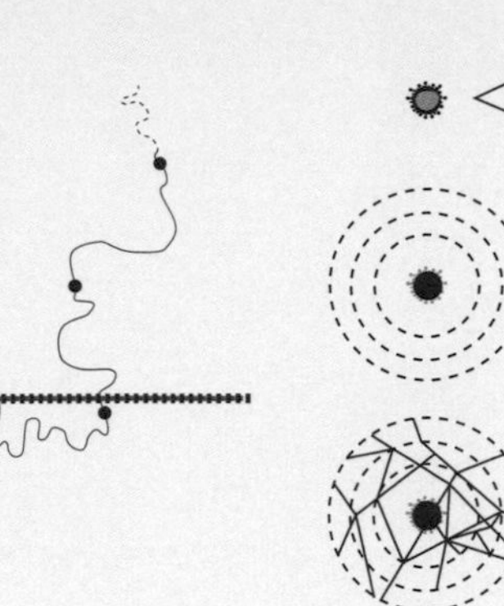

5 *Terra Forma* offers an exploration of the territory with alternative maps created from a redefined GPS point. This point is actually a moving, living point, one that partially generates the space around it.

6 By modifying the space around it, the point, which we call the "point of life," influences and is influenced by other points of life inhabiting the same territory. The Earth is no longer inert matter.

7 Is it still possible to observe this animated Earth and make a map with a distant and disembodied eye? How should space be drawn from within and among these points of life? How can we understand their movements, their interactions, their significance for the geomorphology of landscapes? *Terra Forma*'s experimentation aims to deduce new qualities of space by observing points of life, and to reconstruct models for maps that can organize this material in order to offer it in turn as a tool for the negotiation, sharing, juxtaposition, and regeneration of damaged territories.

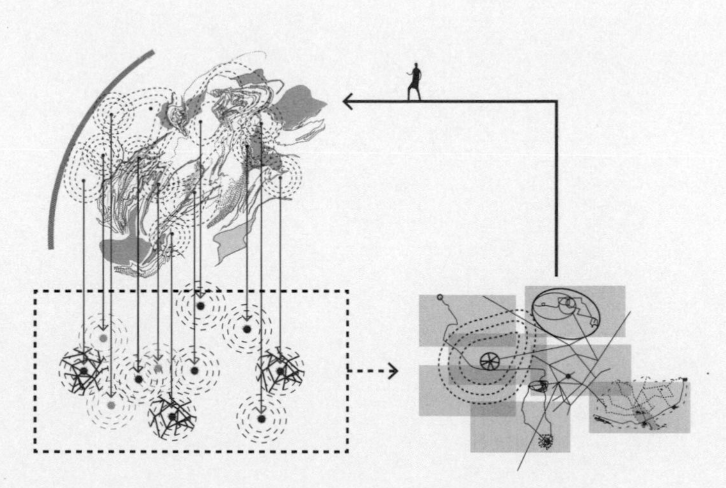

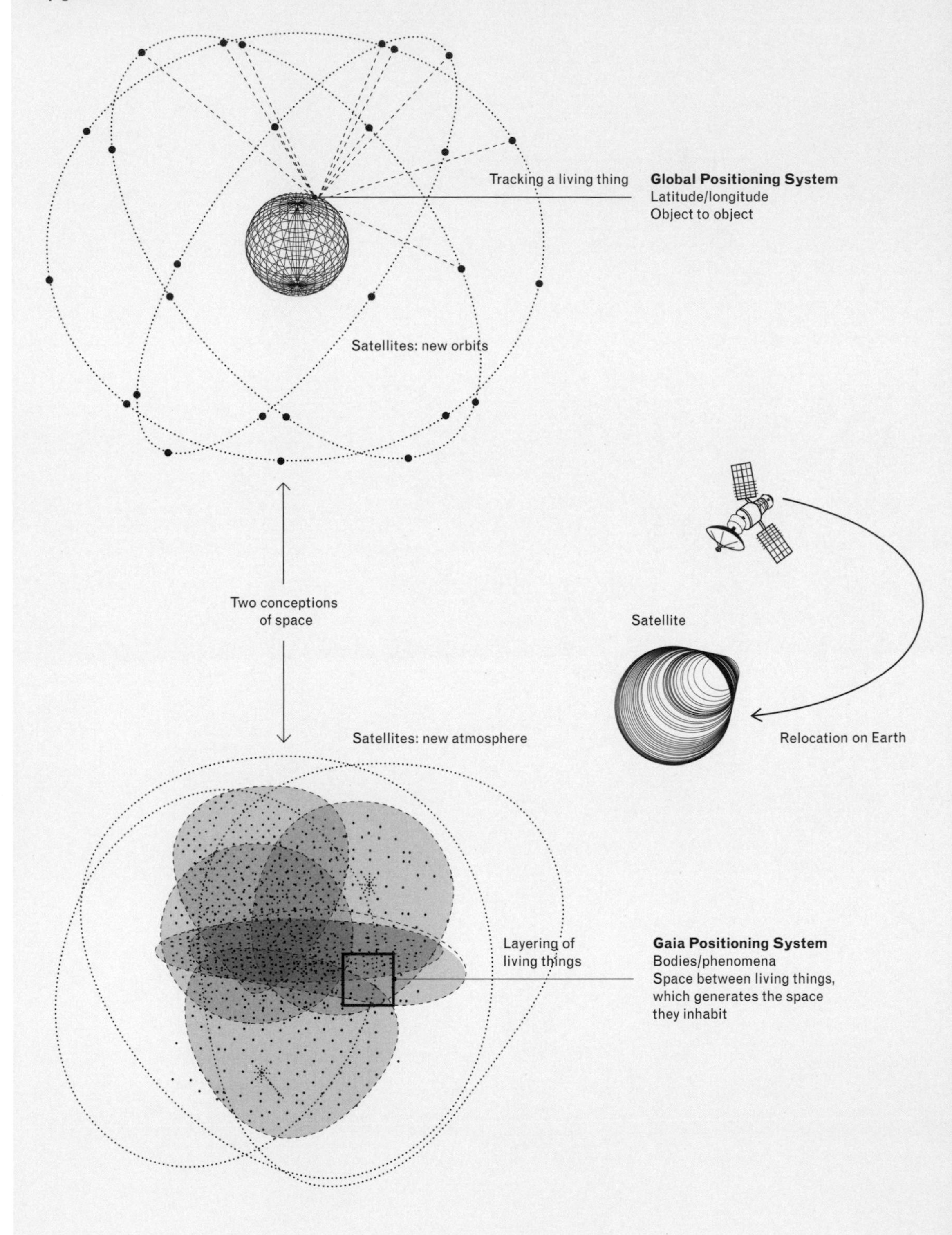

fig. 3 From a GPS point to a living point —a cartographic transformation

framework of metrics and move toward an ecosystem-based conception. The proposed models constitute prospective tools for envisioning not the future of the climate but that of our milieu: what scientists working today in the new Earth System Sciences (which combine geology, biochemistry, and stratigraphy to study the Earth as a system) call the "Critical Zone"—the thin outer layer of the Earth where living things and their resources are concentrated. Taking this new definition of the habitable part of the Earth as a starting point, the following models are perspectives on reality, possible visions of the world sketched through different prisms like so many optical instruments: through depths, movements, points of life, peripheries, pulsations, voids, disappearances, and ruins. They produce situated, embodied knowledge.[3] Each model considers a specific question, a problem that it unfolds in space, sometimes in time, using a borrowed and inflected archetypal figure: the globe is turned inside out like a glove, the grid of coordinates of Euclidian geometry deforms and evolves depending on the actors who cross its space, while the divisions of urban time bend into spirals of space-time.

It was the territories themselves, and the investigations we conducted there, that gave us the material and impetus for these model-tools.[4] The situations we encountered in our practices served as our starting points: these were projects aimed at rehabilitating ruined, abandoned, isolated, polluted, uninhabited sites, for which contracting authorities (generally the public authorities) had asked us to develop reconversion strategies. In these landscaped or urban areas suffering from overpopulation, whose relationship with living things is blocked, it is often up to the architects of our generation to restore conditions of

3 Donna Haraway, "Situated Knowledges: The Science Question in Feminism and the Privilege of Partial Perspective," *Feminist Studies* 14, no. 3 (1988): 575–599.
4 See the appendix for a detailed list of terrains.

habitability—to design habitable worlds. To do so, we had to consider a much deeper notion of ground than we had previously imagined, superimpose life territories, and include the parameter of time. In short, we had to shift our accustomed cartographic and spatial categories, sometimes radically. This book is the result of this constant back and forth between experimental data, attempts at modeling, and visits to sites. Each of the maps presented here is based on a specific territory that we encountered and studied, or on several territories whose salient characteristics we have combined in order to display the model's maximum potential for representation. Applied to different territories, each model could generate different maps.

As Jerry Brotton reminds us, there is no "progress" in cartography toward greater precision or more objectivity (the famous "realism" of Google Earth being just one style among many); rather, cartographies provide diverse cultures with specific visions of the world at precise moments in history.[5] *Terra Forma* reexamines maps but also uses them to examine the state of the world. One of the goals of our experiment is to modify the classic attributes of the map to take into account data and entities that are human and nonhuman, living and nonliving, and which have often been invisible in conventional representations. In its own way, each map reveals the agents who create this open-air spatial stage that is the territory. Considered as a stage, the map becomes a modular, open tool where stories and situations unfold: a new *theatrum mundi* that makes room for natural agents, mingled and associated with the human comedy. But there is no longer any inert, manipulable scenery here that we can rearrange at our leisure. Rather, the maps that follow suggest that it is the actors (human and nonhuman) who make the theater; it is living things who make the world. There is no

5 Jerry Brotton, *A History of the World in Twelve Maps* (London: Allen Lane, 2012), 27.

longer any reason to distinguish what has been kept separate for so long: the theater of humanity and the theater of nature. Humans have long considered themselves to be directors of the theater of nature: builders, shapers of mountains, the sole organizers of space. Without denying this demiurgic aspect (after all, we are creators of space, like all living beings) or this tendency to stage the world, it is clear that the role of humanity has changed: humankind is no longer solely in control, it creates together with many other actors, it makes way for what we call animate entities: human and nonhuman, living and nonliving agents who shape space. The desire to reconceive our poietic and demiurgic relationship to space is what brought us, the authors, together. Repopulating maps amounts to accepting the idea that we humans are not alone in making them.

The following maps are "living maps" rather than "maps of the living," not only because they are always under construction, always moving, but because they try to capture and integrate the part of production that is carried out by living things, rather than proceeding from the belief that the human point of view should predominate. Of course, the mind immediately jumps to the benefit of making dynamic digital versions: living maps in the sense of animated, interactive, evolving maps. If we imagine developing these tools a step further, alongside other media and formats (such as models and workshops), offering an atlas of "living maps" wouldn't have to be limited to animating classic maps (placing living things on an already established map), but could also make living things the main agents of the map. This is why, in our eyes, this book is an essential step in the project. It emphasizes, in contrast with the maps we know, the changing status of the map, which is no longer the result of patient topographical plotting, but a tool of documentation that captures the transformations of a territory produced by biotic and abiotic agents. The role of the architect-cartographer thus undergoes a striking transformation, less concerned with developing a territory than with choreographing the movements of living things. The builder-architect thought in terms of visibility, symbols, landmarks, and markers; they wanted to mark space, create a monument, organize space in order to organize people. The architect-choreographer is interested in the circulation of living things; they see ruins as extraordinary reservoirs of uses to be invented, and don't conceive of space beyond the forms of life that constitute it, perceive it, and produce it. The role of the cartographer is similarly disrupted. At the outset of this adventure, we wanted to map the world in a new way by multiplying the tools used to do so. But we gradually discovered a method of visual representation that rendered any single overarching grasp of the material obsolete. During the course of our exploration, worlds became more complicated, and the static tool of the map became potential, unfinished cartography, offering instead provisional model-tools.

Written in the style of a tale of exploration, this book also aims to be a light, portable drawing manual for testing techniques on various terrains in order to progressively and collectively constitute a new kind of atlas. We use the term exploration in the spirit of Robert Hooke, English scientist and member of the Royal Society, who undertook to rediscover the world around him in 1665, a world already thought to be well understood, by means of his optical instruments.[6] Through the lens of his microscope, he revealed magnifications of the most commonplace items: the structure of cork, the magnificence of insects, landscapes of mold. An entire, minuscule, unknown world, which he called *terra incognita* in reference to the Age of

---

6 *Micrographia* by Robert Hooke is one of the founding texts of experimental science, published in London in 1665. The discovery of a "new New World," *Micrographia* was one of our graphic, methodological, and narrative models in writing this book.

Discovery. Like Hooke, we are not discovering new countries, but we are learning to see the territories around us differently: the Chemical Valley near Lyon, a part of the Pyrenees, Paris and its suburbs, as well as the 10th arrondissement. Like Hooke, we have explored them using new tools for visual representation—the seven models presented in the book's seven chapters. We describe their techniques for production, in the hopes that others will take them up and improve on them. Hence the space dedicated to sharing the sketching gestures that allowed us to construct these models and maps. The first gesture is a circle, because we had intuitions about feedback from our actions as inhabitants of the world. The idea was, implicitly, to move from the notion of an infinite space to the consciousness of a limited world (in terms of space and resources). The circle is also the lens Hooke used to discover the microscopic world, the table for a board game where conflicts and alliances play out, and the rotation of a thought in motion.[7] The second gesture is an inversion. This movement causes the horizon, which had once seemed distant and unattainable, to curve to surround us, so that everything we try to send far away returns to us (Soil model). The map is the territory; space is no longer a mere receptacle for living things but rather the result of their actions (Point of Life model).

We realize that most of these maps are not easy to read if you don't first accustom yourself to their codes and legends. In a similar way, an Indigenous map will remain inscrutable unless you are informed by locals about its meaning. A Western map is made precisely for dispensing with Indigenous people and their ways of directing attention toward a manifold landscape. To opt for indecipherable maps without a guide is to suggest a different system of discovery and proximity. One might object that these

7 The circle is precisely one of the recurring iconographic figures in *Micrographia*.

maps, devoid of topography and coordinates, are no longer maps. Though they may not resemble the maps we are familiar with, they nonetheless retain one of their objectives: to provide orientation. Many think that orientation means isolating stable seas and immobile physical landmarks so that we can get our bearings and triangulate a location. But when the ground itself seems to shift, landmarks cannot remain the same; the fixed point around which everything turns is the activity of living things themselves. This landmark is no longer immobile like a mountain peak, but if we train our attention intently on its transformations, it remains the only reliable option for grasping our position and our place in landscapes that are now alive.[8] To orient ourselves here will thus be an attempt to inhabit a space populated by other living things, other entities who share and shape the Earth with us, *terraforming* it.[9]

Just as we sometimes speak of speculative narration, we could speak here of speculative visualization: how can we return the power of being seen to these other living things, these agents and actors of the Earth?[10] In this effort to revisualize worlds, we are the inheritors of a long history of Western cartography. However, our objective is no longer to transform the Earth, but to produce frameworks through which the Earth's entities can inscribe their traces. By means of science or practical knowledge, paying attention to events, clues, and manifestations of these entities, we "document" them in models designed

8 Mountain peaks are themselves no longer fixed points because of climate change.
9 Terraforming is generally understood to be ultratechnical and human—such as the hypothetical terraforming of the planet Mars, which would supposedly transform its atmosphere and climate to make it inhabitable and colonizable by humans. Here, in contrast, we define *terraforming* as an eminently terrestrial practice shared with other living things.
10 See in particular the work of the École de recherche graphique and Fabrizio Terranova; works by Didier Debaise; Donna Harraway on SF; and Aliocha Imhoff, Kantuta Quiró, and Camille de Toledo, *Les Potentiels du temps: Art et politique* (Paris: Manuella éditions, 2016).

to welcome them and make them visible.[11] It is a matter of recording them so that we know how to engage with them in conversation, of constructing a process of cosmopolitical notation using old and new tools.

This remaking of tools is chronicled in the Snark maps located at the beginning of each chapter. Borrowed from Lewis Carroll, they indicate the invisible, unrepresentable course traced on the sea: the old landmarks are useless when you're going on a Snark hunt.[12] These blank maps pose the question of the disappearance of the background map and offer a place to try out a new way to orient oneself on the sea of the world. In other words, they are a call to walk and to draw, as two symmetrical or simultaneous movements.[13] Each Snark page shows, side by side, the tools of a classic map and the shift that the model attempts to make: where we are coming from and where we are going. As we move through the models, we gather tools that allow us to draft maps, establishing a new frame of reference step by step. This book recounts the episodes of this quest in seven stages leading to the discovery of the world, the discovery

---

11 We have inherited the "arts of noticing" from Anna Tsing, at once the art of observation, the art of noticing, and the art of documenting and tracing. Anna Tsing, *The Mushroom at the End of the World: On the Possibility of Life in Capitalist Ruins* (Princeton: Princeton University Press, 2015).

12 He had bought a large map representing the sea,
Without the least vestige of land:
And the crew were much pleased when they found it to be
A map they could all understand.
"What's the good of Mercator's North Poles and Equators,
Tropics, Zones, and Meridian Lines?"
So the Bellman would cry: and the crew would reply
"They are merely conventional signs!
Other maps are such shapes, with their islands
and capes! But we've got our brave Captain to thank:
(So the crew would protest) "that he's bought us the best—
A perfect and absolute blank!"
Lewis Carroll, *The Hunting of the Snark* (London: Macmillan, 1876), 15–16.

13 Francesco Careri, *Walkscapes. La marche comme pratique esthétique* (Arles: Actes Sud, 2013).

of the limits of its exploitation and the infinite possibilities of its exploration—because the soil is infinite in its stratification, and the curvature of its surroundings is infinite in its movement. Could the Earth be discovered anew by adopting a centrifugal rather than centripetal mode of thought for its exploration? Could our picture of it be renewed? We hope so, and it is our goal in writing and sketching this book to contribute to the vast movement to renew the Earth's cosmograms.[14]

14 The term *cosmograms*, in particular as it is used in the history of science and in anthropology by John Tresh, from whom we borrowed it, designates material, graphic, iconographic, or architectural devices that represent a cosmology and, in our case, a certain conception of the earth.

fig. 4 The human ground

The layering of parcel maps from different eras from an industrial complex in Marl (Germany), or the modern palimpsest of the appropriation of land.

MODEL I

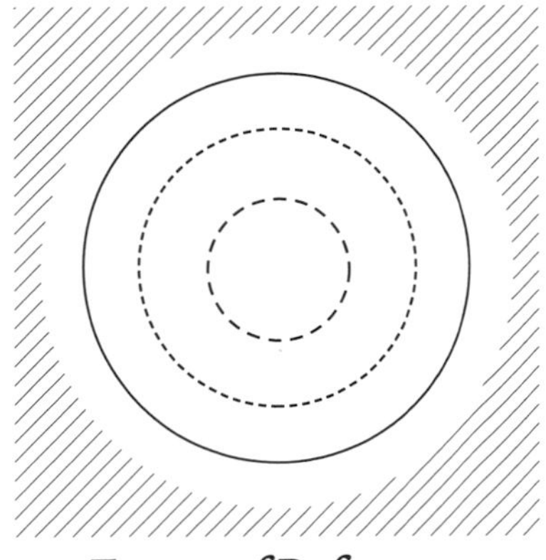

Frame of Reference

SOIL

Orientation

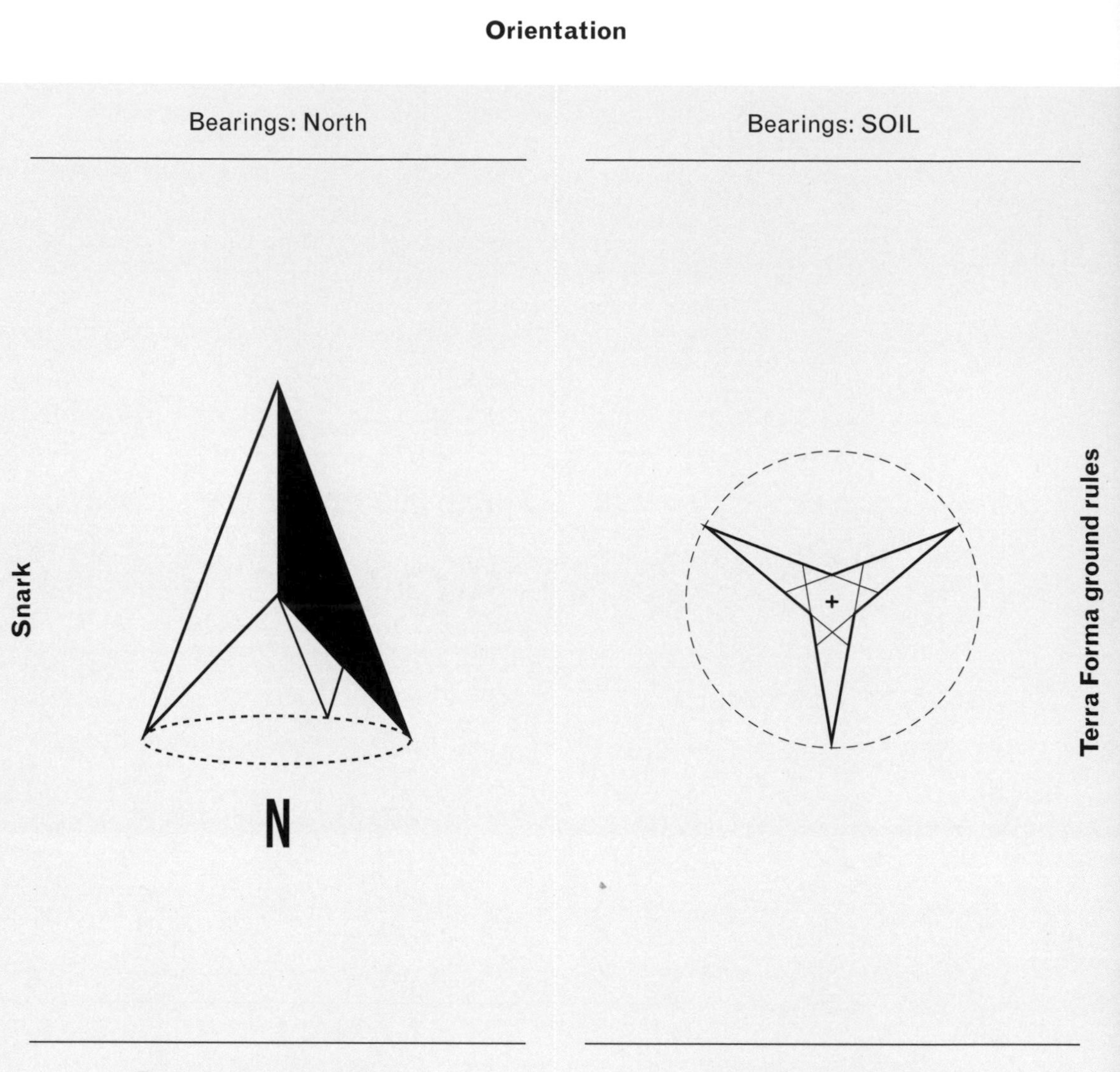

SOIL

When using a map, the first frame of reference we rely on is direction. For a long time, our maps indicated which way was north—magnetic north, but also the Northern Hemisphere centered around Europe. However, another much more ambiguous frame of reference bursts onto the scene with the appearance of Gaia.[15] This frame of reference is at once more powerful and more fragile, more global and more localized, more worrisome and more reassuring. It unites communities as much as it divides them. This "new-old" frame of reference is the ground.[16]

But from the first day of our expedition, the ground gave way beneath our feet. We thought we were beginning our journey with a confident step, traversing a solid and stable terrain at a brisk pace, but it turns out to be heterogeneous, fragile, and active. Scientists who study the Earth system are the first to warn us: topography is no longer a stable frame of reference, because the Earth itself isn't stable; it is a historical being that hasn't been inhabited and habitable for very long, and above all, its continents are still moving. So we must abandon the very idea of a fixed territory that is delimited once and for all. This is why we will dig deeper before traversing it. The ground is always represented as an aerial map, but we are looking for a way to view a cross section of it as a thin surface, or from below. The idea of observing the globe from the inside isn't new, but earlier experiments focused mainly on the surface, allowing the visitor to climb inside a hollow model, as in the georamas popular in the nineteenth century.[17] We are more interested in understanding its thickness, strata, and layers; a cartography of lower layers of soil, underground. A map of the Earth's underside rather than its surface.

15 Isabelle Stengers, *In Catastrophic Times: Resisting the Coming Barbarism*, trans. Andrew Goffey (Open Humanities Press, 2015).

16 Bruno Latour, *Down to Earth: Politics in the New Climatic Regime*, trans. Catherine Porter (Cambridge: Polity, 2018).

17 Yann Rocher, ed., *Globes: Architecture et sciences explorent le monde* (Paris: Norma, 2017). The work of tremendous erudition in this catalog and the eponymous exhibition allows us to identify historical antecedents to this understanding of the ground as thickness rather than surface: the description of an "infinitely thin layer to which human existence is confined" (20), the idea of subcutaneous geography (172), or even a three-dimensional conception of the environment (222).

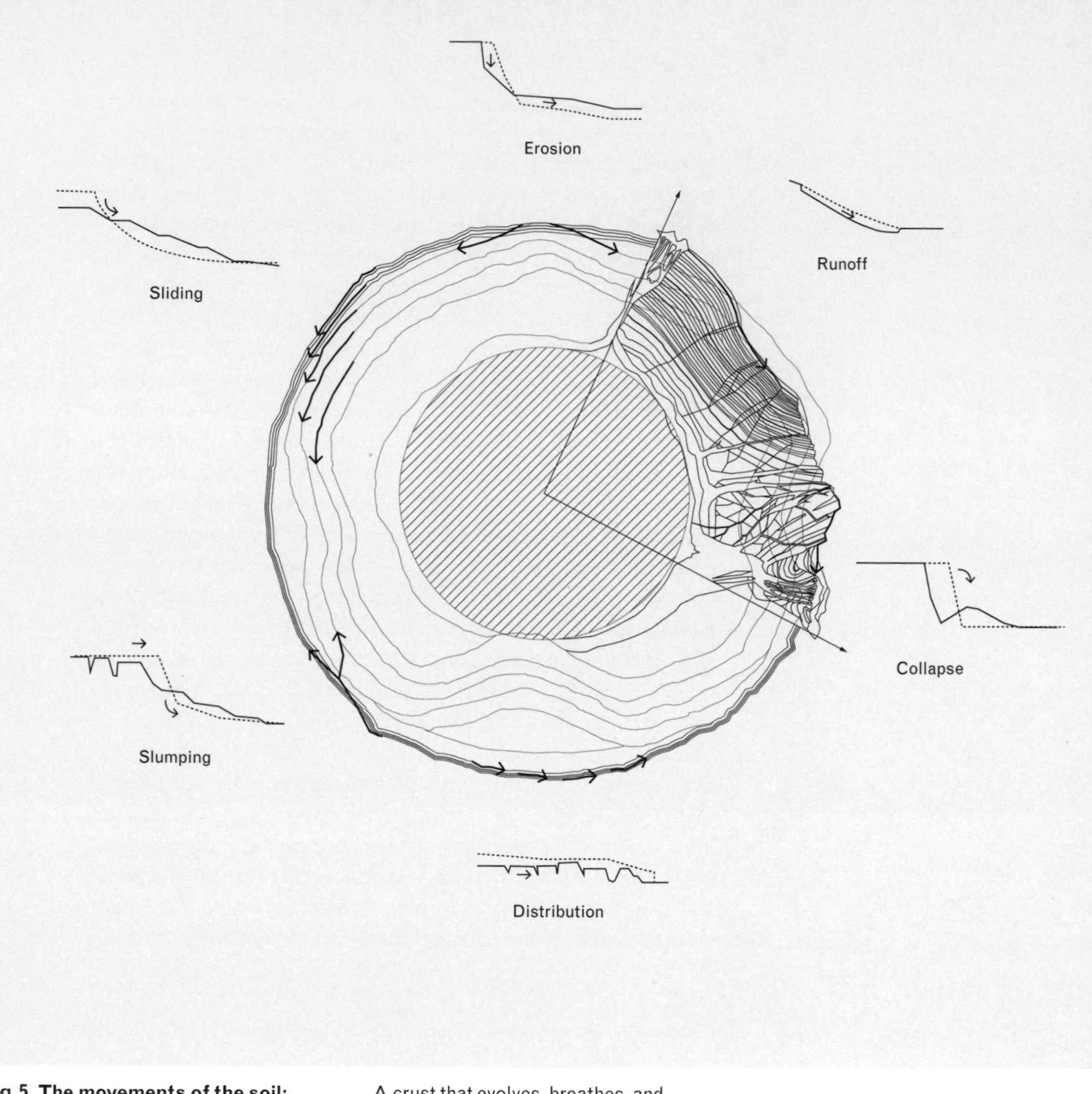

fig. 5 The movements of the soil: active phenomena.

A crust that evolves, breathes, and transforms.

Our land registries represent the ground as linear, abstract slices, attributing to certain humans deeds of property that grant them ownership of a piece of ground, and sometimes even the area below the ground (fig. 4). Multiple reconfigurations of parcels throughout history—carving up, consolidating—bear witness to the socioeconomic conflicts linked to the appropriation of land, very often with dramatic consequences for the population, both human and nonhuman. Even today, some planners rely only on this parcel economy to design their projects, thus directing the lives of the populations in the area. Zoning maps dictate how cities are generated, but they also do so for the countryside. Some urban historians build urban theories based on this long history and orient their projects toward a larger consideration of earlier layouts, reading the land like a parchment of superimposed traces, a palimpsest. For example, when we speak of the "urban fabric," we envision the city as an organic whole, which generates questions about continuity and discontinuity. When we construct a building, should it exist as an individual entity, signaling discontinuity, or should it disappear into the continuum of the urban body? Should we build vertically according to traces in the ground and archaeology, or not?[18]

From the point of view of geologists, however, the soil appears quite different. Subject to multiple entropic and negentropic pressures, it shifts, crumbles, slides, sinks, and collapse (fig. 5). Its waves spread both horizontally and vertically. The motions of the tectonic plates and the water cycle driven by the sun's energy, combined with the actions of living organisms, mean that the ground is never at rest, from the Earth's crust to the thinnest layers of the fertile surface. From local pressures to unalterable depths, geological history is divided into eras that span

18 Aldo Rossi, *The Architecture of the City*, trans. Diane Ghirard and Joan Ockman (Cambridge, MA: MIT Press, 1982).

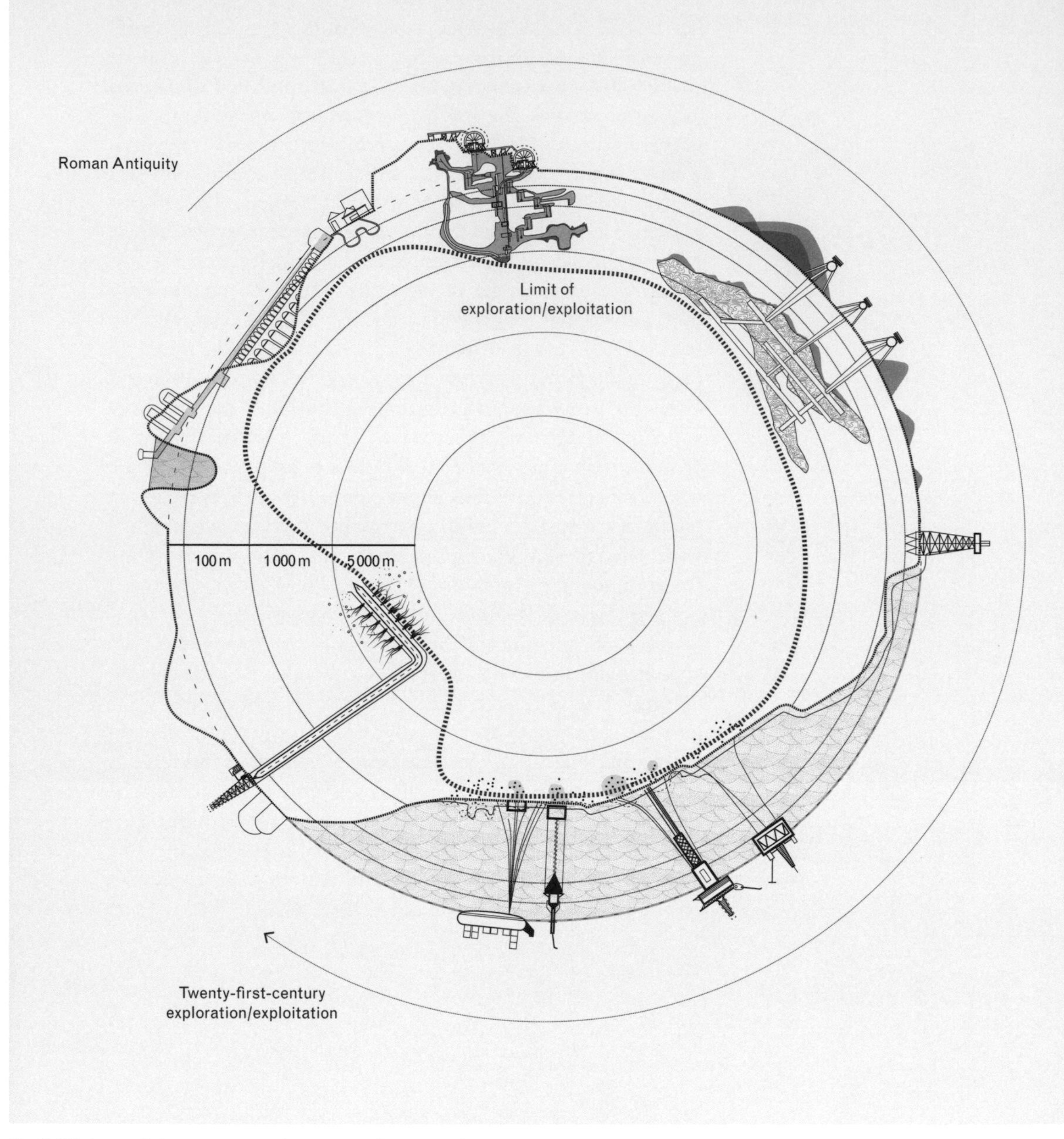

fig. 6 History of the conquest of the ground by human beings

From wells for accessing potable water (from 1 to 100 meters in depth) to shale gas extraction (from 1,500 to 3,000 meters in depth), with oil wells in between.

millennia, from which humanity seemed to be excluded until recently. But twenty-first-century geological history is different. In the face of data collected by scientists from around the world[19] and the work of historians,[20] we can no longer deny or remain indifferent to the fact that humanity has become a major geological force, rapidly transforming not only the chemical and physical composition of the Earth, but also disrupting ecosystems at a pace that gives them no time to adapt. The actions of humanity exploit the Earth's depths and disrupt the temporal strata, creating strata of plastic and concrete, crafting the Earth as if it were a machine we own (fig. 6).[21] Humanity has become an "Earth-eating" entity, to borrow a striking image from Deborah Danovski and Eduardo Viveiros de Castro.[22]

To explore such complex terrain, we must first create the right tool, a sort of telescope for viewing the interior of the Earth. Whereas Galileo directed his telescope toward the heavens, we will point ours at the soil. Our expedition will be vertical (fig. 7). From the outset, this first model-tool requires a certain flexibility on the reader's part. By means of a thought experiment, we find ourselves at the core of an inverted globe, one that has been turned inside out like a glove. Its exterior, the atmosphere, is at the center, suddenly confined in a closed, reduced, constricted space. Its deepest part is now arranged in concentric circles

---

19 See, e.g., Jan Zalasiewicz et al., "Petrifying Earth Process: The Stratigraphic Imprint of Key Earth System Parameters in the Anthropocene," *Theory, Culture & Society* 34, nos. 2–3 (2017): 83–104.

20 Christophe Bonneuil and Jean-Baptiste Fressoz, *L'Événement anthropocène: La Terre, l'histoire et nous* (Paris: Points, 2016).

21 Jan Zalasiewicz, "The Extraordinary Strata of the Anthropocene," in *Environmental Humanities: Voices from the Anthropocene*, ed. Serpil Oppermann and Serenella Iovino (Lanham, MD: Rowman & Littlefield, 2016).

22 Deborah Danowski and Eduardo Viveiros de Castro, *The Ends of the World*, trans. Rodrigo Nunes (Cambridge: Polity, 2017), 77. The authors adapt the expression "white earth eaters" from the work of Davi Kopenawa and Bruce Albert, *The Falling Sky: Words of a Yanomami Shaman*, trans. Nicholas Elliott and Alison Dundy (Cambridge, MA: Harvard University Press, 2013).

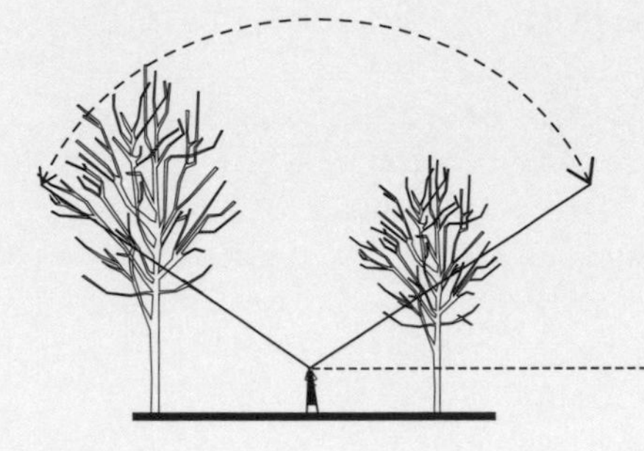

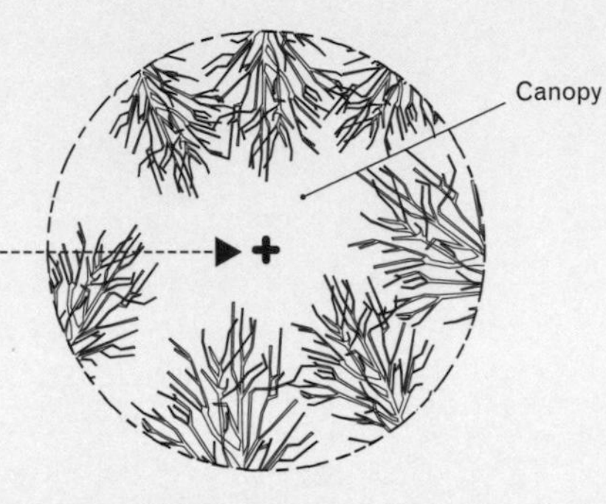

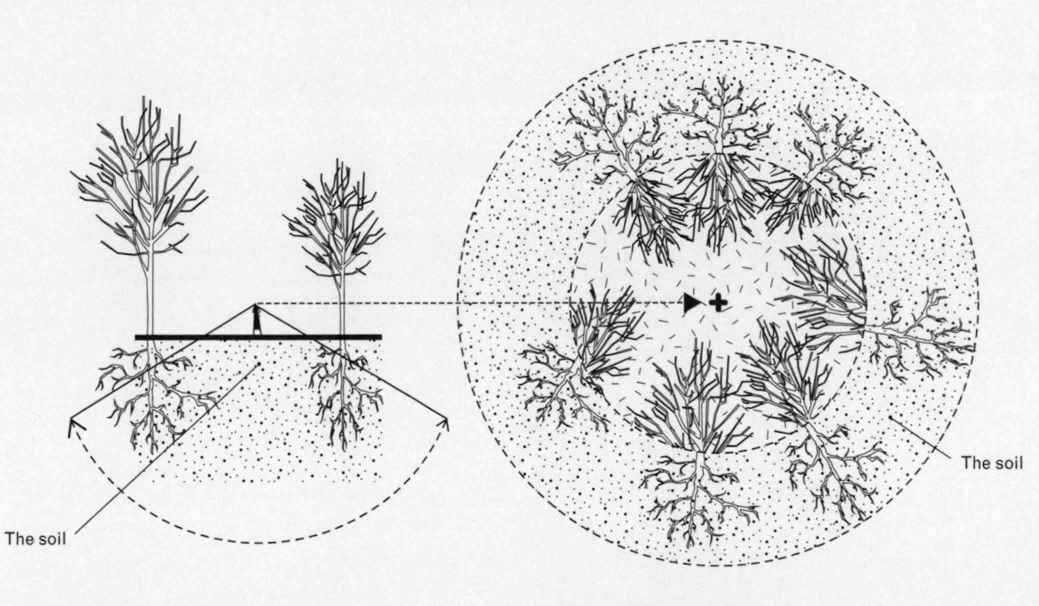

fig. 7 Visualization of the vertical strata from the canopy to rocks, from many "points of view" at ground level

Reorganization into "site plan and cross-section," which allows the reader to grasp the dimensions of the Critical Zone in terms of depth and surface.

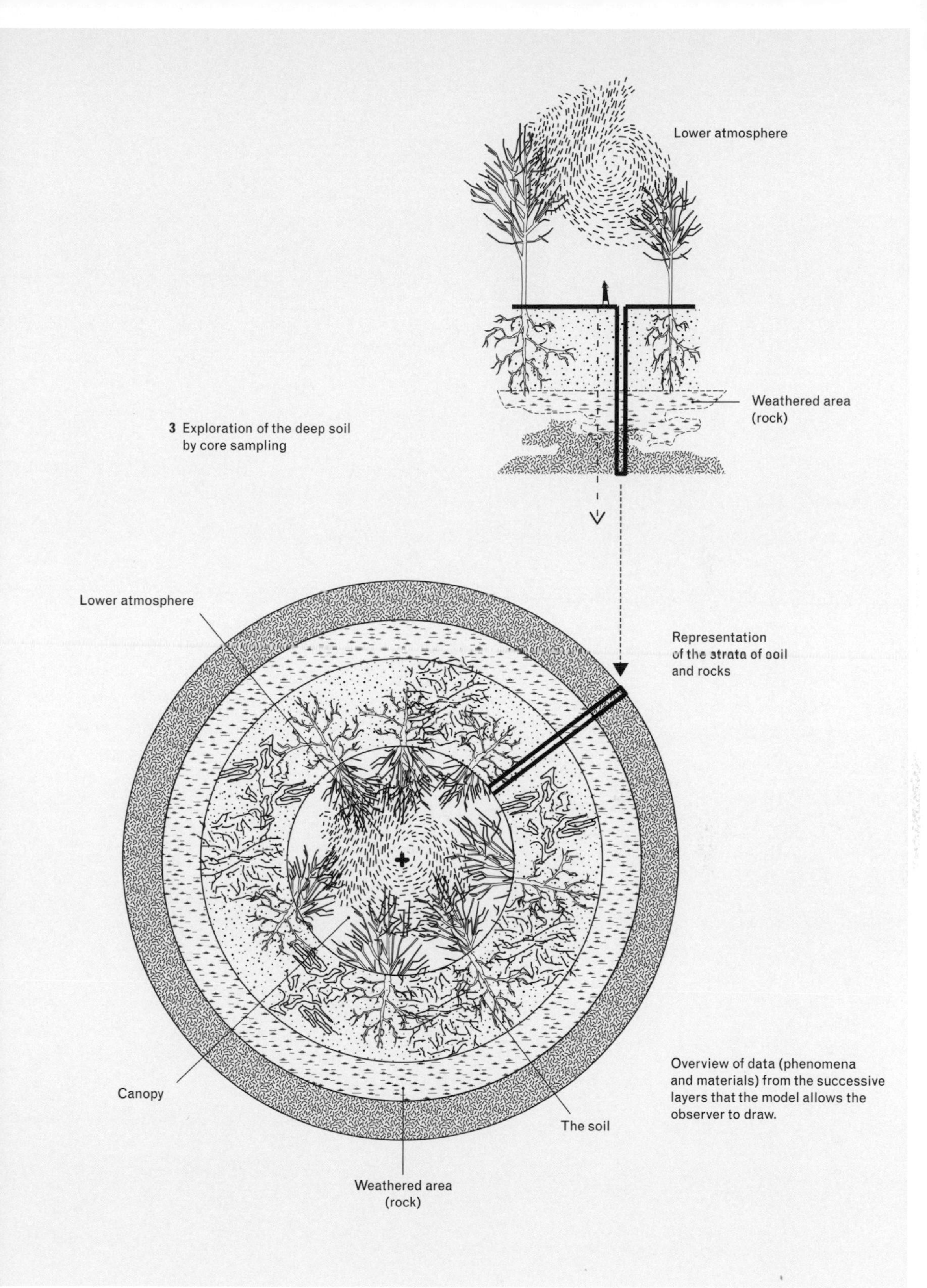
Lower atmosphere
Weathered area
(rock)
3 Exploration of the deep soil
by core sampling
Lower atmosphere
Representation
of the strata of soil
and rocks
Canopy
Overview of data (phenomena
and materials) from the successive
layers that the model allows the
observer to draw.
The soil
Weathered area
(rock)

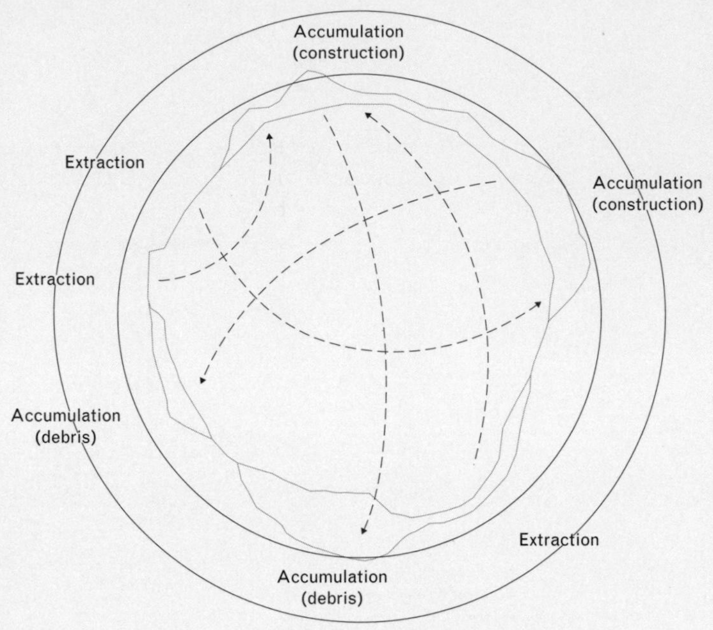

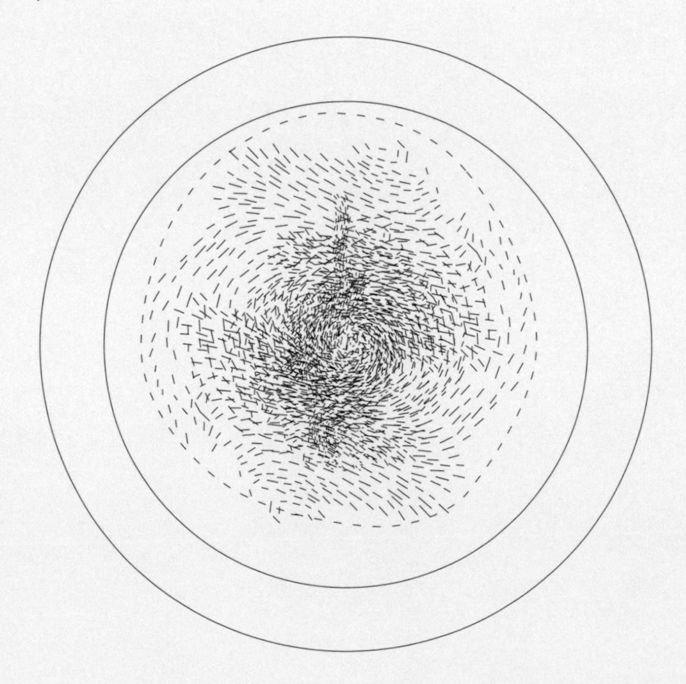

fig. 8 Boomerang effects

reaching the outer edges of the map. In this way, the entire map is focused on this Critical Zone, the thin film of Earth where water, soil, rocks, and the living world interact, and where human and nonhuman life are concentrated, along with their resources (see model pages).

This new frame of reference has many consequences. For one, it makes the Earth visible as a palimpsest by bringing to light the stratification of land use. The reach of human influence is legible not only horizontally but also vertically, through the accumulation of human actions that have penetrated the soil to the deepest strata. The model makes it possible to visually represent the constant flux between different levels of the ground—a fluidity rarely associated with the Earth. Here, a landslide has caused two strata to meet—two epochs that never should have met. Elsewhere, water causes toxic substances to infiltrate, contaminating deep groundwater (see map pages). The second consequence of this model is that it makes tangible the boomerang effect (fig. 8). By isolating the atmosphere at the center, it emphasizes the fact that everything we disperse into the atmosphere returns to us because we live in a closed system. Everything happens as if nothing were "outside of the world." Even the central atmosphere is full: nanoparticles, clouds, airplanes, birds, a dome of pollution. Each object released into the sky necessarily falls back to the ground, sometimes rebounding. At its depths, the available space is no longer infinite, because the lower layers of soil are already saturated. Turned inside out, the soil suddenly reveals a different cosmology, a buried architecture, its organism inhabitants. The soil, which had seemed homogeneous and solid, is moving, porous, and liquid when viewed from below; it is composed of particles and agents that constantly imprint and modify its structure and form. These optics make visible the Earth's skin, along with the many agents who create it, signaling the nearly equal portions of human and nonhuman production that make up this material (fig. 9).

In fact, biogeochemistry teaches us that the border between animate and inanimate material is a convention of naturalist classification. Lynn Margulis explains that "the vital bond between the environment of the Earth and the organisms upon it makes it virtually impossible even for biologists to give a concise definition of the difference between living and nonliving substance."[23] This is why we plan to embark on a voyage of exploration of the Earth as earth—soil, humus—rather than as a globe. The Soil model is subsequently used to analyze this Critical Zone, the film around the surface of the Earth where life-forms are distributed from the rocks to the canopy.[24]

To test this first model, let's first consider a well-known archetypal industrial site: Chemical Valley south of Lyon along the Rhône, home to oil refinieries that are being converted to green chemistry sites (see map pages). The industries are located on an 800-hectare area of backfilled soil extracted from the river. The original morphology of the river had been disrupted to create a canal in order to simplify the commercial exchange of materials. The soil on which the industrial facilities are situated is therefore nearly totally inert and has been deeply polluted by the various businesses that have operated there for years. The territory is full of pipelines that link the buildings to each other (for exchanging liquids or gases via surface or underground pipes). These facilities stand against the backdrop of numerous natural spaces connected to the river. The diversity of the area and its soil—half natural, half artificial—makes it an ideal object of study.

Now let's drive the model into the ground. Down 1 meter . . . 5 meters . . . 10 meters . . . 50 meters . . . 100 meters. Beyond this point, our ability to see weakens, then is snuffed

23 Lynn Margulis and Dorion Sagan, *Microcosmos: Four Billion Years of Microbial Evolution* (Berkeley: University of California Press, 1986), 72.

24 See Alexandra Arènes, Bruno Latour, and Jérôme Gaillardet, "Giving Depth to the Surface—An Exercise in the Gaia-graphy of Critical Zones," *The Anthropocene Review*, vol. 5, no. 2, 2018, 120–135.

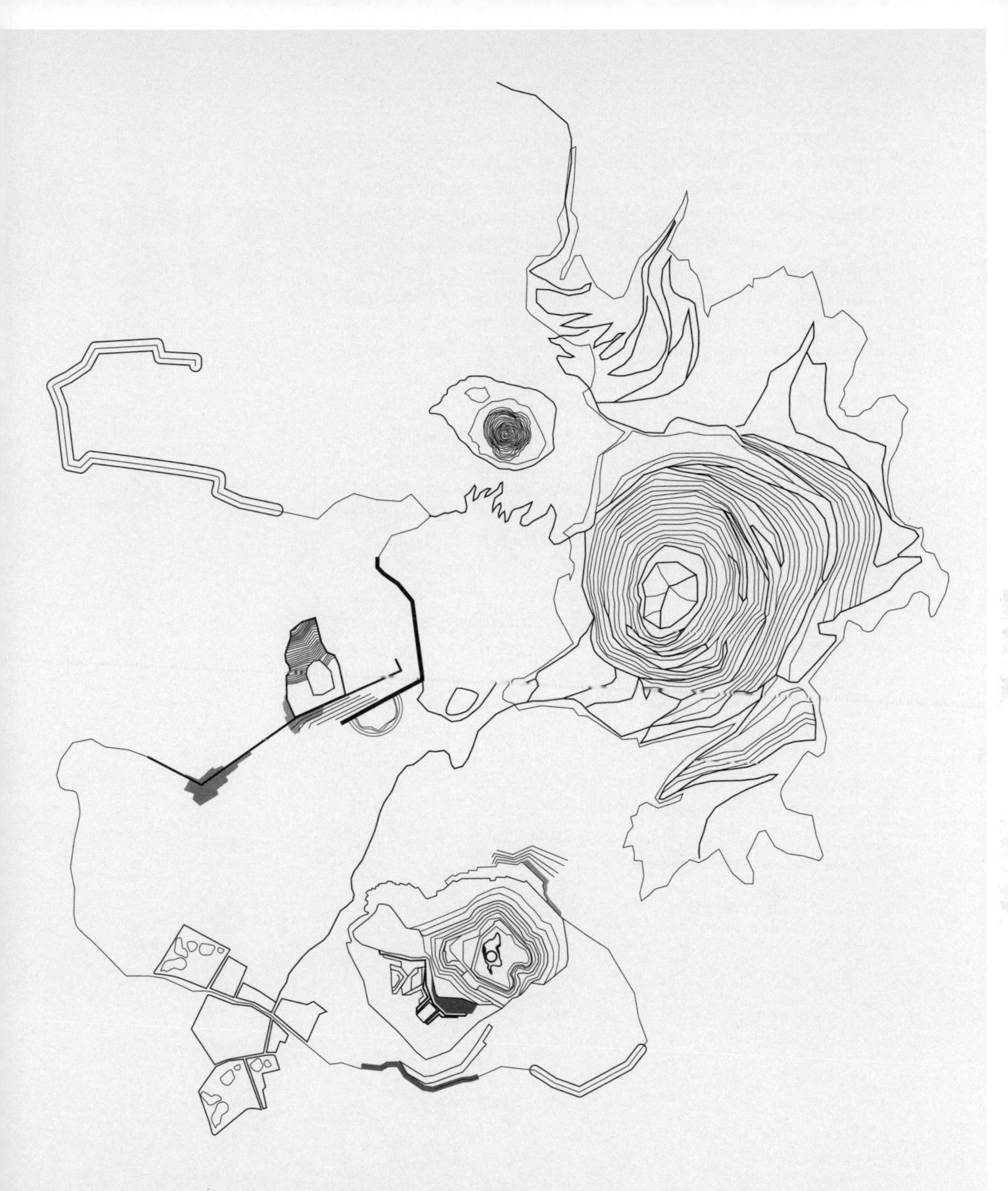

fig. 9 Territorial tattooing

Mines are characteristic examples of the act of "terraforming" performed by humans on their environment. The consequences of this transformation of the terrestrial environment are the extraction of resources, displacement of earth, and modification of the horizon of the ground. These alterations call the meaning of habitability into question.

out entirely. In the first meter, the ring of the model closest to our epicenter extracts a material that swarms with movement and is composed of a multitude of organic particles; the surface, having been turned over at a depth of just 1 meter, is lined with micro-fissures and weighs a few thousand tons. But in certain places, we find nothing at all; the ground is dead. It isn't empty, but nothing is moving. Instead, pieces of debris, more or less whole, sink deeper. At –5 meters, we follow mole tunnels and the burrows of small mammals, which guide us back toward the surface. The tunnels are interrupted here and there by a concrete sarcophagus, tanks connected by what looks like a network of pipelines. Between the tanks, the water seeps and drains their contents deeper. At –10 meters, we discover the vestiges of former buildings, large and rectangular. They tamp down the ground and mix with the rocks. In the other hemisphere, roots have already cast them back up to the surface, as if they weren't welcome in the subterranean world. However, for the most part, the organism inhabitants and the hosted objects coexist in the first levels. At –50 meters it is no longer possible to distinguish the origins of one from the other; chemistry performs its transformations on organisms and inert matter alike. At –100 meters, rock bears witness to the material's final state. The model allows us to represent these layers visually by doing away with scale and scanning the strata horizontally, then resituating them concentrically around an empty space, the round center, our atmosphere (see model pages).[25] A fossil can be as visible in it as a mine, a volcano as visible as a tank, a network of pipelines as visible as the water that flows through the rocks—because on the map's surface, their effects have become equally significant. The phenomena present at the site being studied are divided among the

---

25 To test the limits of this model, we added indicative or revealing elements to the Chemical Valley site: a volcano, shale gas extraction, and so on. The cartography presented is thus not referential but "potential," constructed on a composite model.

quarters of the map by uniting opposites: extraction/creation of new fossils (northwest); seeping/absorption (northeast); concretion/explosion (southeast); digging/geologic uplift (southwest). The soil, like an inverted skin, reveals what is happening directly below its surface, which makes it live or die. This is our *terra incognita*: the soil beneath our feet.

Underneath the human-drawn parcels of land, we have tried to reestablish the alterity of the soil, its inalienability because organisms not only inhabit it but shape it. The Soil model shifts our understanding away from maps as tools for grabbing land and annexing living resources (later we will see, in the (Re)Sources model, how this annexation functions and how we might imagine a different model), and which thereby make the Earth an inert surface, even though soil is densely populated. Consequently, if we can no longer say that we define ourselves by possessing a certain territory (a practice that now seems archaic, given how dynamic life on Earth is, along with the growth of our human population), how can we define ourselves today while respecting the alterities of each being (both human and nonhuman alterity)? The Soil model opens a breach, a rift in how we used to define a territory and the globe, and thus in how we define ourselves. The globe cannot float in suspension in the air; in the model it is caught and seems flattened by gravity. The successive layers that constitute the soil and its various depths are spread out and smoothed, like a globe spread on the ground, showing the complex and heterogeneous landscape of the Earth's layers and no longer presenting the illusion of a world that is crystalline and aerial (see transformation pages). The entities made visible on the map of the soil grip us in a common history, intermingled in the Earth's jumble.

However, we can make out each entity, each component on the map. Each one distributes itself differently and reacts to other entities closer or farther from it that act on its living space.

Thus, if we can discern each one's actions and space, we can draw a map of them. This map would be a prerequisite for any political negotiation between these entities, allowing territory to be defined, a dwelling space, a living territory, in order to then bring them together with the living spaces of others and to make it possible to layer them.[26] What principles can we agree on? How can we generate a common landscape? This task awaits us in the next two models. Model II, Point of Life, endeavors to describe living territory for each of us, starting from a modified center that configures the rest of the map. Model III, Living Landscapes, tackles the description of landscapes that are animated, rustling, and entangled.

---

26 Baptiste Morizot, *Les Diplomates: Cohabiter avec les loups sur une nouvelle carte du vivant* (Marseille: Wildproject Éditions, 2016).

MODEL I
SOIL

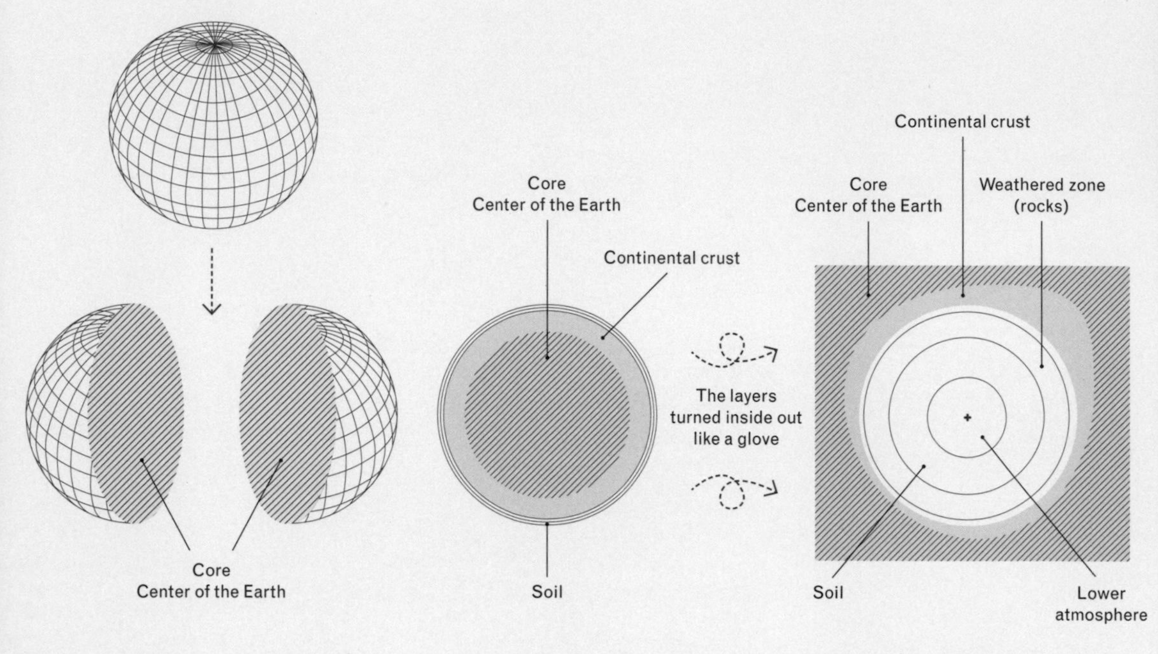

1
Split the globe down the middle to obtain two equal parts

2
Lay one of these cross sections across the surface of a plane. In the resulting map, the rocky core occupies the center and nearly the entirety of the diagram, while the soil and the atmosphere are nearly invisible around the exterior circumference of the globe.

To remedy this difficulty (the invisibility of creatures' living space), carry out the following inversion: reverse the order of the concentric circles from the core to the atmosphere. On the basis of this manipulation, or anamorphosis, the rocky core is spread to the edges of the map, while the atmosphere (the sky and the air that we breathe) is found at the center of the diagram.

Continue the same operation with the other conventional strata and geological layers. Starting from the edges, we have the following order: core, crust, weathered area (rock subject to the actions of the sun and water, where traces of biological life can be found); soil (our humus, fertile soil, and the soil we live on); and the lower atmosphere (from the soil to the tree canopy). Some scientists working in the earth sciences call the strata from the weathered area to the canopy the "Critical Zone."* We also use this term to designate the layers that make up the new frame of reference soil.

*The "Critical Zone" is the thin, superficial film around the surface of the Earth where water, soil, rocks, and the living world interact. Geochemists call this zone "critical" because this is where life, human activity, and resources are concentrated.

Base model

Rock layer
is more significant

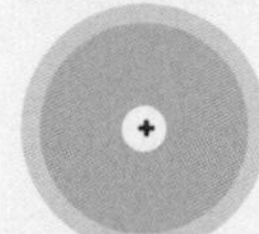

Soil is more significant

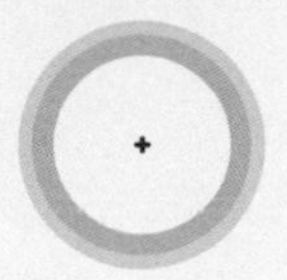

Lower atmosphere
is more significant

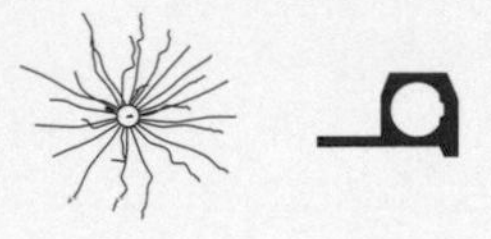

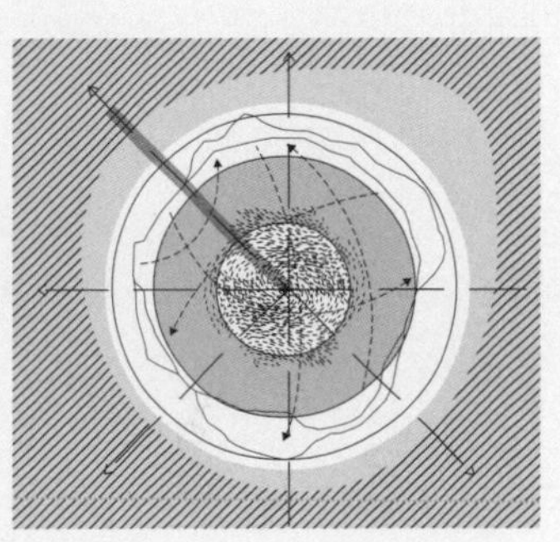

3

Adapt the diagram to the chosen site of observation and define the depth gradient to suit the geological structure of the site, the presence of a thicker or thinner layer of soil, topographical relief, and so on. (The positioning and circumference of the circles are thus parameters based on the inherent qualities of each site being studied). The circles of depth simultaneously represent a site plan and a cross section (viewed from below or from the interior, with these profiles juxtaposed on the site plan according to a depth gradient).

4

Record the composition of the soil being observed according to the two proposed categories: the organisms inhabiting it and the objects hosted in it (see captions on the following pages), whether of human or nonhuman origin. These creatures and artifacts, ignoring the rules of the smooth surface, dig, ingest and digest, dissolve or compact, rearrange and adapt fragments of the soil to make it correspond to their vital needs.

5

Note the physical and chemical composition of the soil, the materiality of the entities present, and their movements. Identify possible disturbances, struggles, or conflicts in the soil.

MODEL I
SOIL

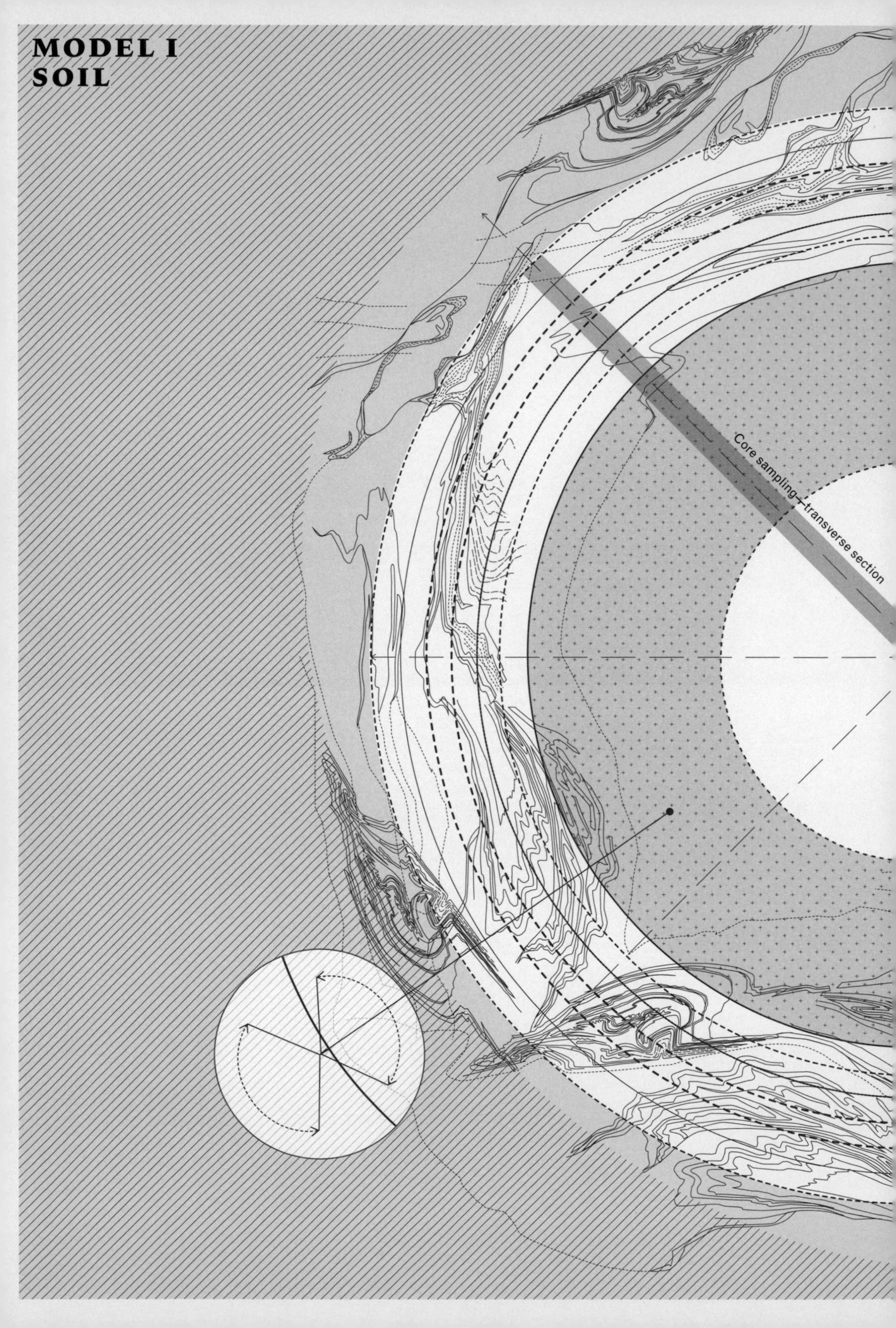

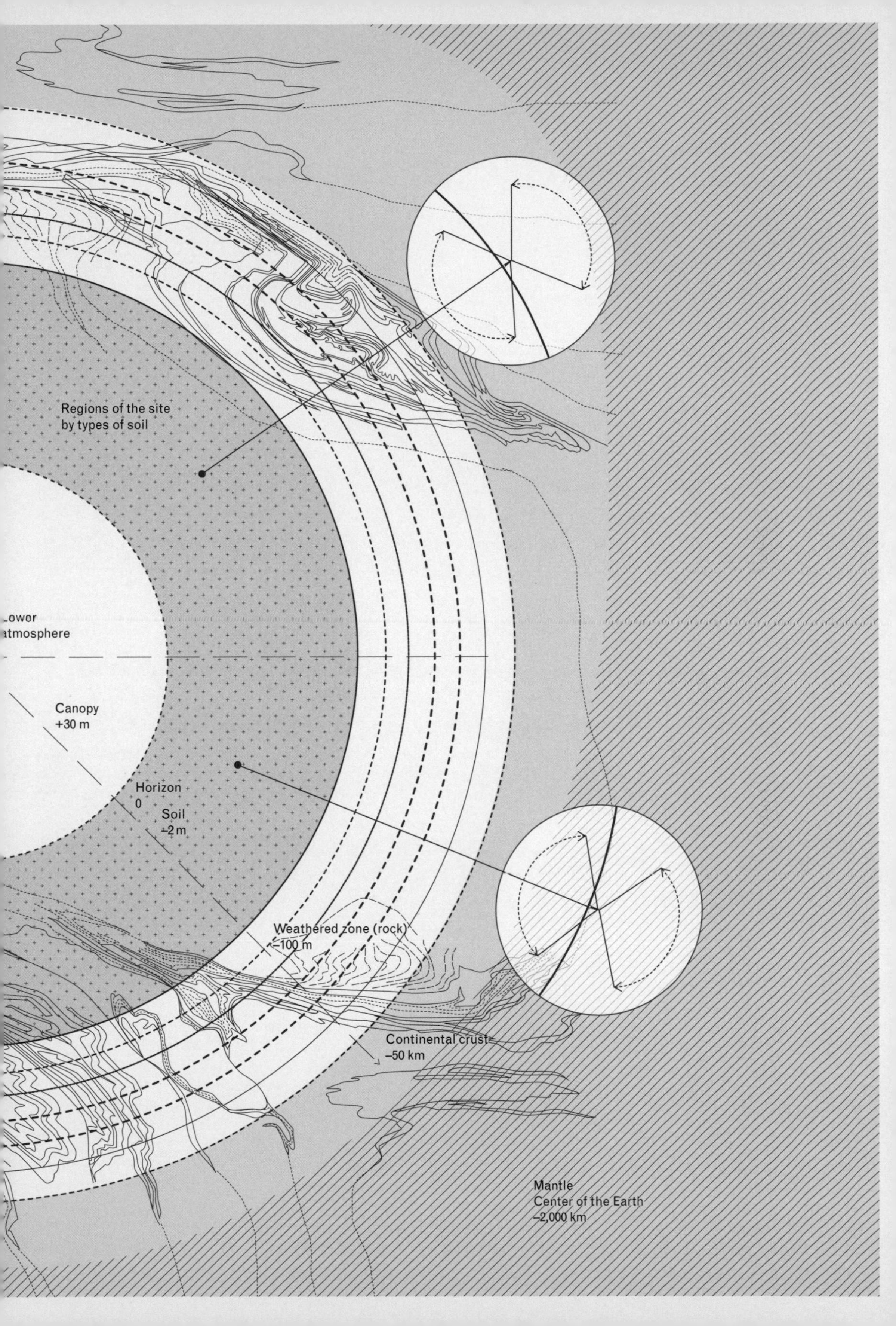

Regions of the site
by types of soil
Lower
atmosphere
Canopy
+30 m
Horizon
0
Soil
–2 m
Weathered zone (rock)
–100 m
Continental crust
–50 km
Mantle
Center of the Earth
–2,000 km

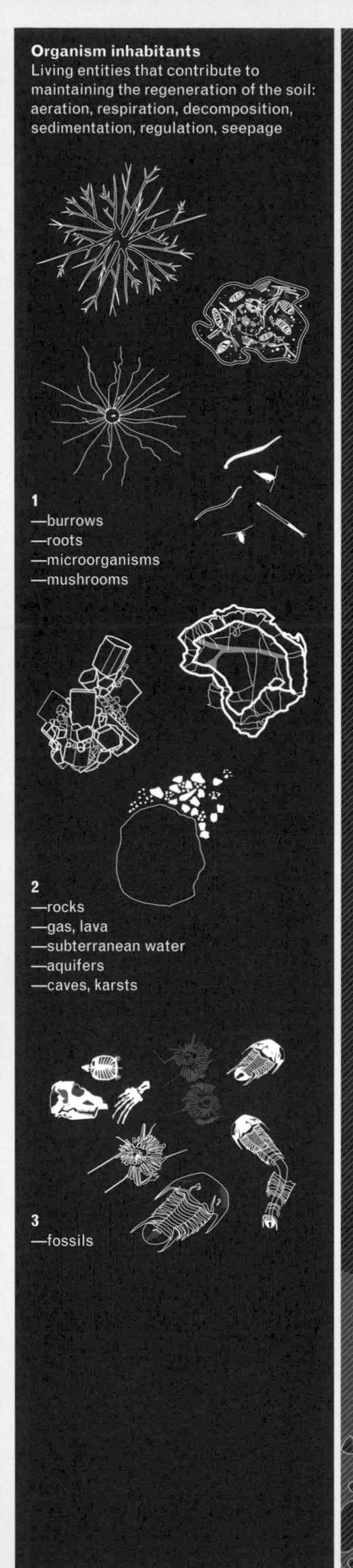

Organism inhabitants
Living entities that contribute to maintaining the regeneration of the soil: aeration, respiration, decomposition, sedimentation, regulation, seepage
1
—burrows
—roots
—microorganisms
—mushrooms
2
—rocks
—gas, lava
—subterranean water
—aquifers
—caves, karsts
3
—fossils

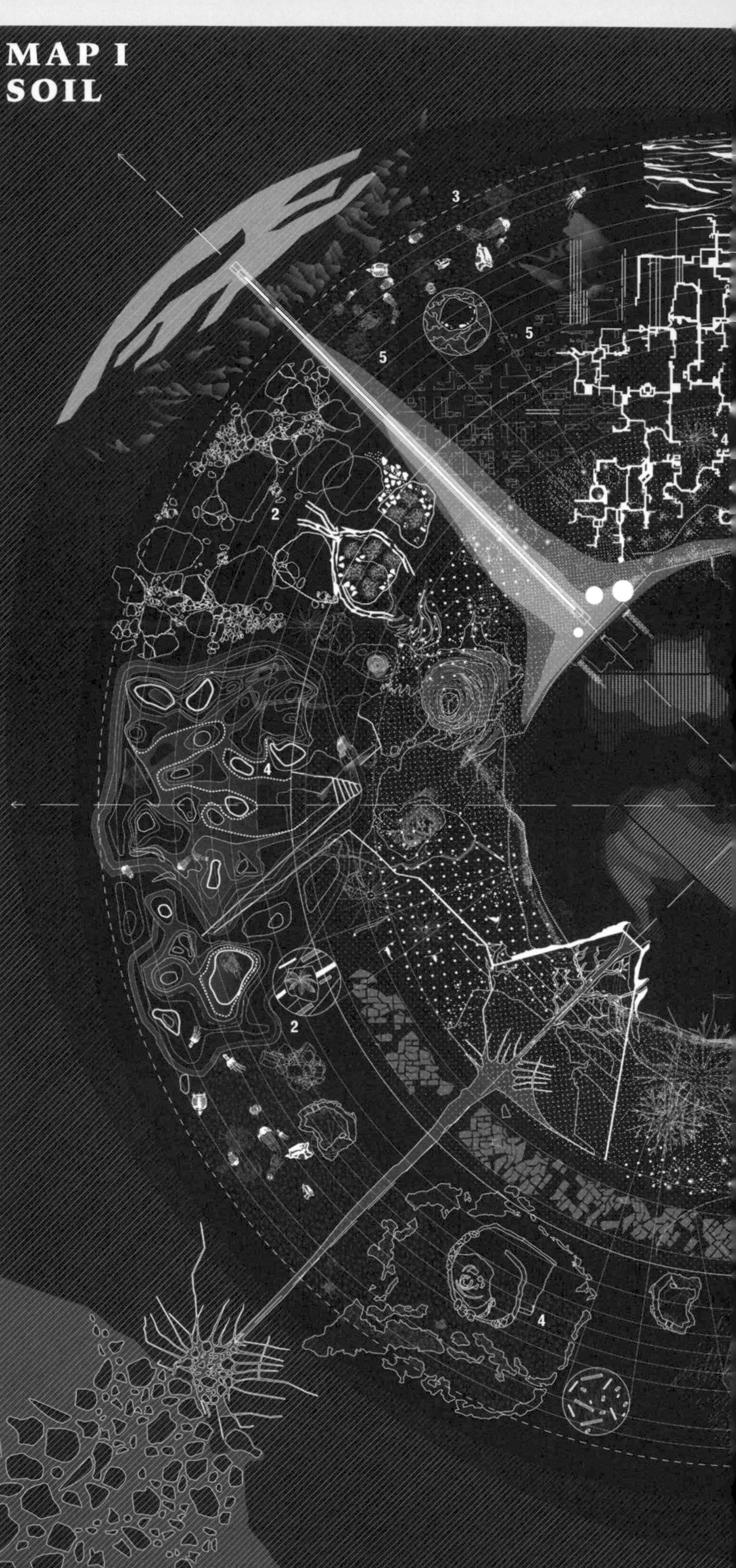

MAP I
SOIL
3
5
5
2
4
2
4

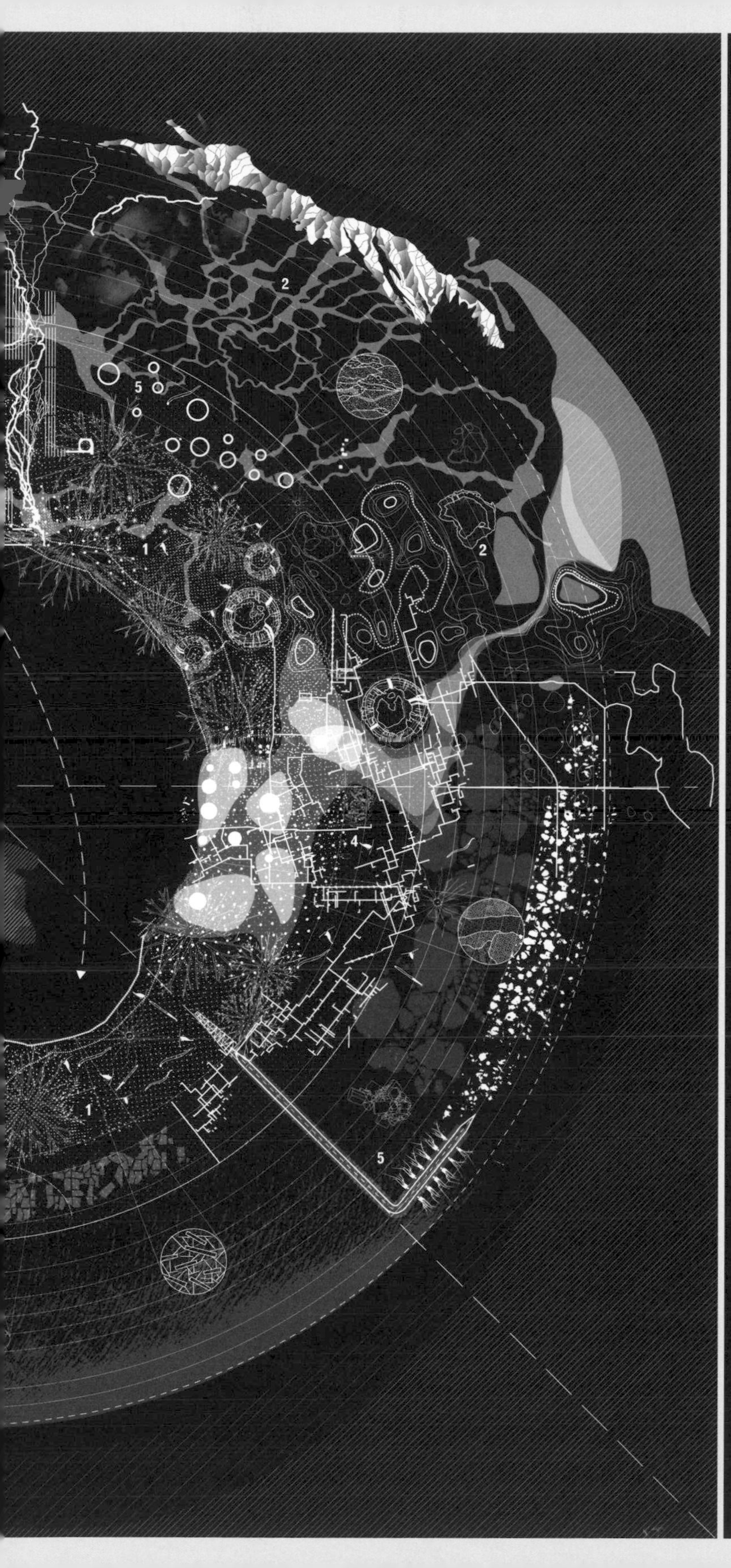

Objects hosted in the soil
Inert elements that alter the composition of the living ground and figure in a more real depiction of the anthropized soils of the Anthropocene.

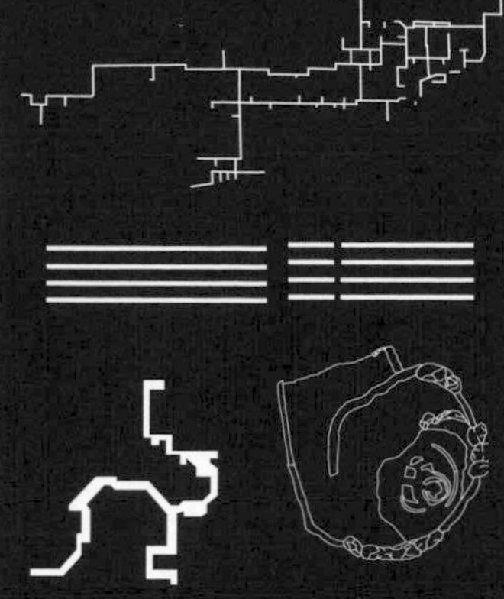

4
—network of pipelines
—various kinds of pollution and contamination of the soil with hydrocarbons
—stone vestiges of the first humans

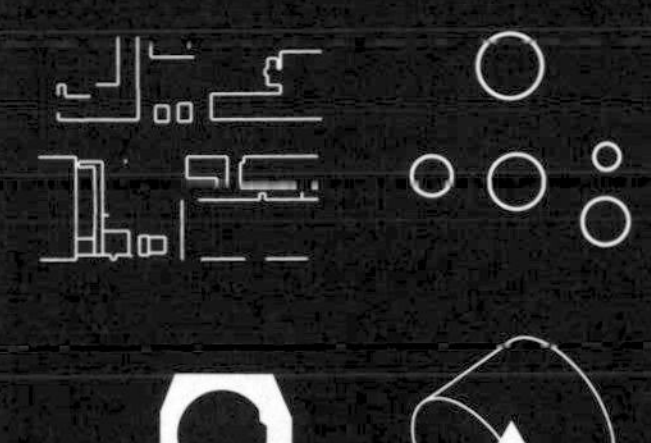

5
—extraction of gas and oil, drilling
—buried nuclear waste
—storage tanks
—tunnels
—mines
—holes
—foundations

TRANSFORMATION I
SOIL

From the globe . . .

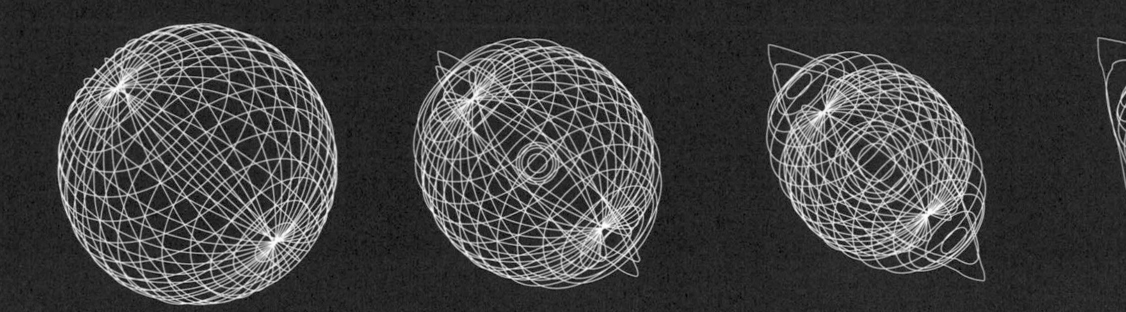

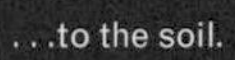

. . .to the soil.

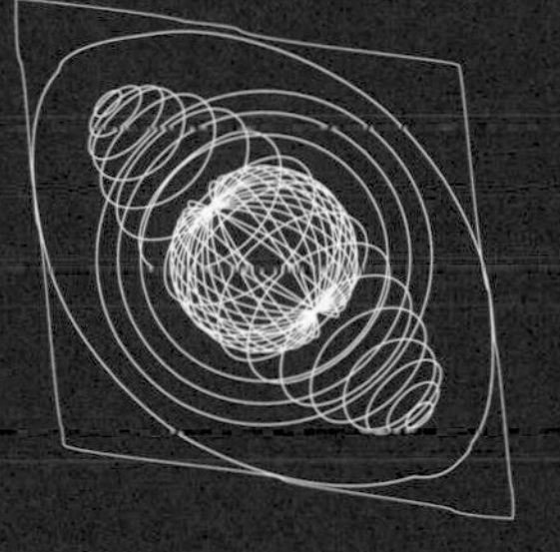

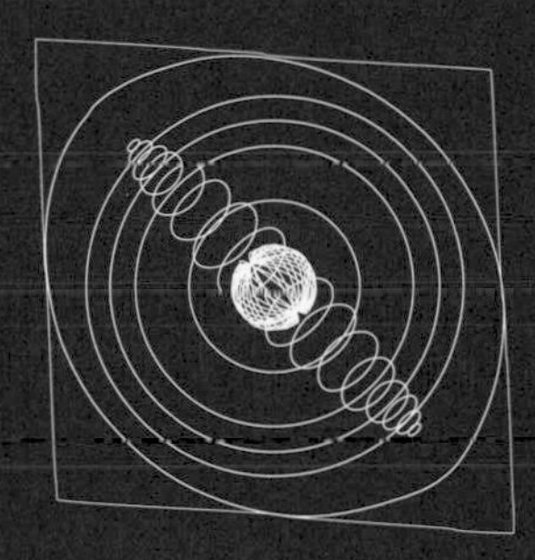

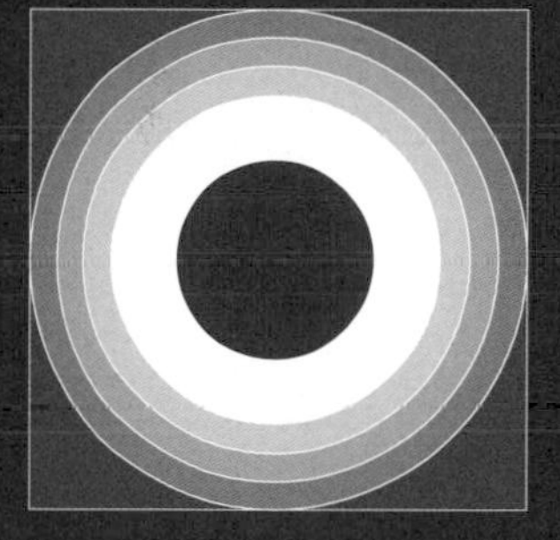

fig. 10 The case of a residential area (Marne-la-Vallée): a succession of thresholds, from a person's door to their neighboring communities

MODEL II

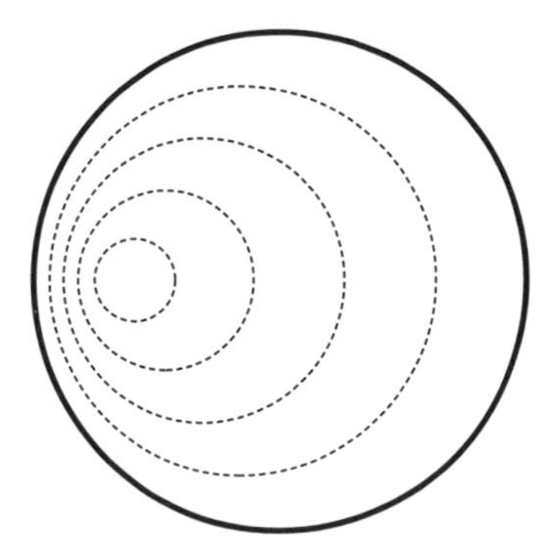

POINT OF LIFE

Positioning

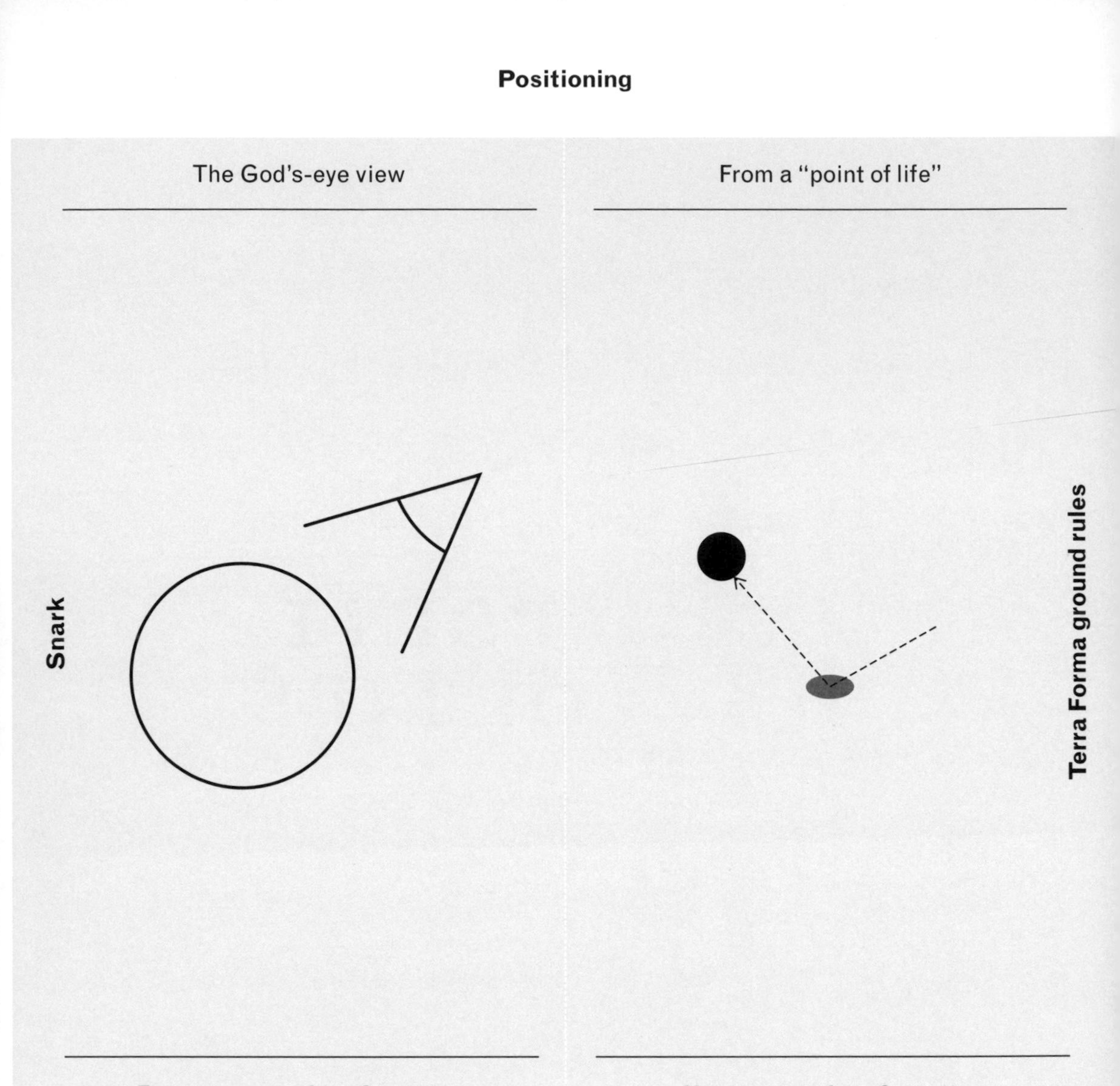

POINT OF LIFE

What is this new ccenter that configurations the map around it? From what perspective should we define our territory? After having abandoned the God's-eye view, the aerial point of view that hovers in zero gravity and globalizes our vision of the Earth, we tried to recover the gravity and material of the Earth with the Soil model, identifying the dynamic activities carried out by organisms that live or once lived on its surface and have now burrowed into its depths. The Soil model plotted a teeming space (full of living organisms and objects that are the subjects of history and memory) rather than an empty one; it revealed a resonant material that is constantly recomposed with each movement, rather than an unobstructed surface to be traversed or conquered. So all we have to do now is reclaim an Earth-bound point of view. The second model is an attempt to represent the world from an animate body, a living point or "point of life" (a powerful formulation by Emanuele Coccia), in order to try to sketch a map of active body-spaces—no space without a body, and no body without space.[27] The point of life thus calls into question the GPS point that we are now accustomed to seeing moving on the map (fig. 10). But what is hidden behind this point located with the help of a globalized system? What is it really made of? How is it anchored to the ground? How does it move? These questions are explored in this model and the following one, Living Landscapes.

These maps entail tracking these animate beings, their movements, traces, rhythms, and affects—qualities that were once called "secondary," which allowed them to be erased from

27 Emanuele Coccia, *The Life of Plants: A Metaphysics of Mixture*, trans. Dylan J. Montanari (Cambridge: Polity, 2017). (Coccia's original formulation, *point de vie*, sounds pleasingly similar to *point de vue* ["point of view"]. I follow his translator, as well as other English-speakers, who have , rendered it as "point of life."—Trans.)

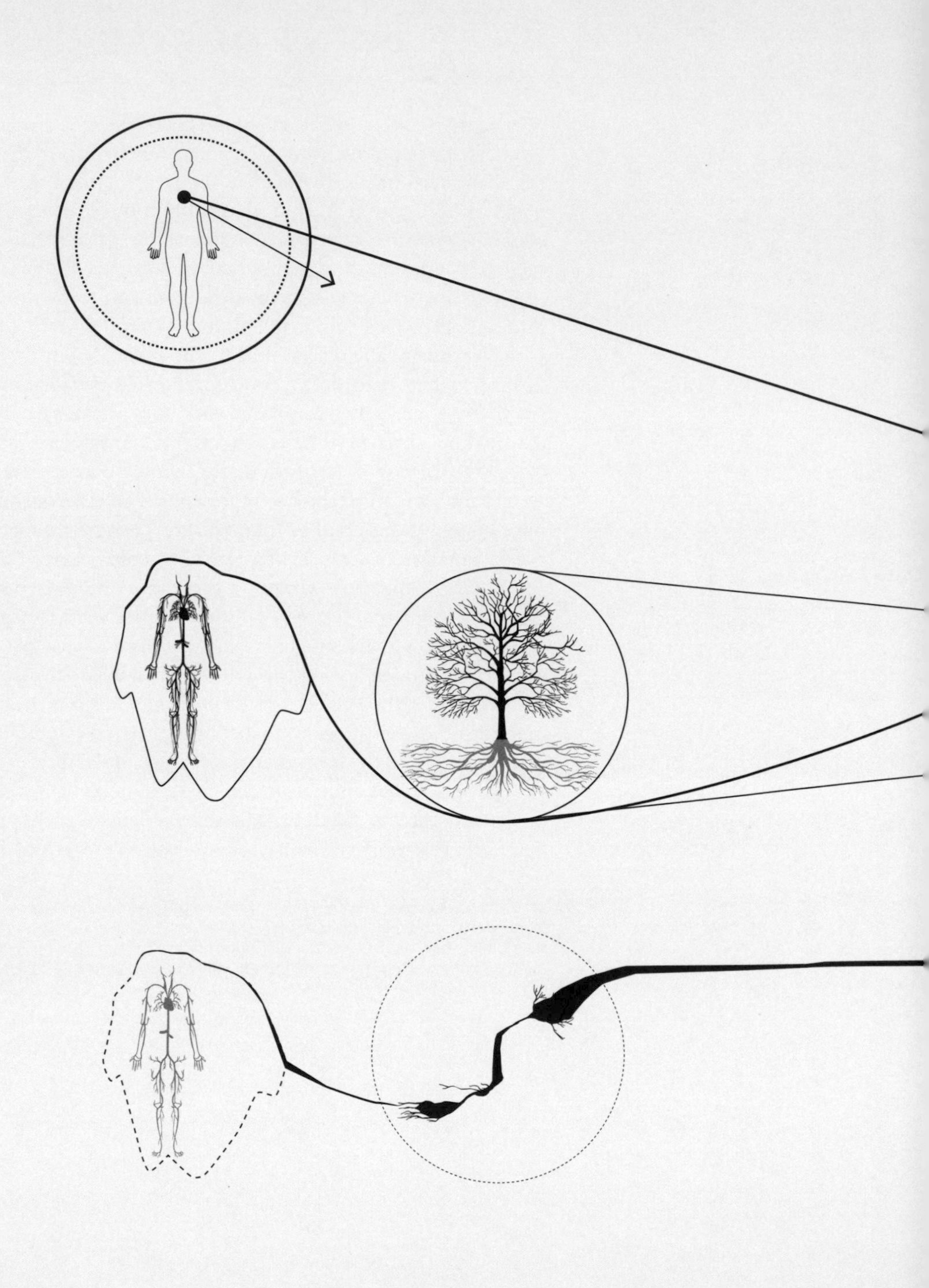

fig. 11 The redefinition of the point of life as an extension of the world

The concept of the point of life invites us to reinterrogate the body's relationship to the world. Being anchored in the world entails the phenomenon of embodiment, which forces us to conceive of the skin, the tree's bark, and the Earth's crust as a continuum of envelopes. Cartography thus allows us to represent these interconnected bodies, from the point of life to the world-space.

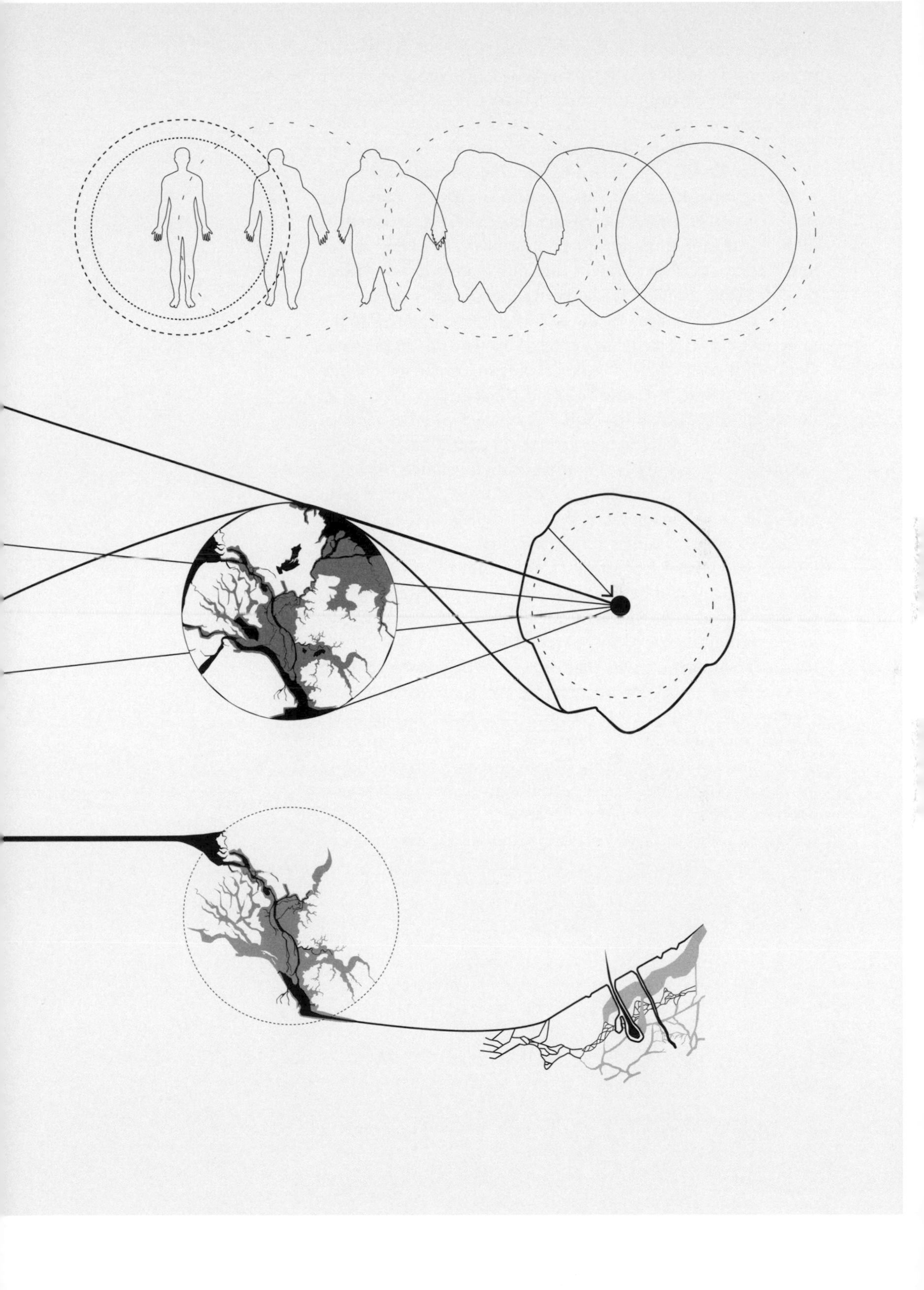

the map, dismissing them from the modern project of quantifying the world and localizing things based on geometrical space. In fact, entities of the living world that are represented on maps lose a great number of their characteristics, notably their potential for growth. On maps, objects are measured once and for all. This gave rise to the standard that an object drawn on paper would be reproduced without alteration, with the same measurements. In contrast, the approach chosen here is interested in living things and thus gives them priority. In the maps that follow, we have tried to reimport into our representations these potential dimensions that have been suppressed.

The redefinition of the soil as thickness and as living material forced us to redefine our relationship to it, our mooring. One of the best examples of a living being that is anchored in the ground is a tree. Imagine an optical instrument capable of unfolding all of its qualities; note everything the instrument has made visible: its potential for growth (in depth, height, size—volume and branching), respiration (carbon balance), flows (of sap), growth (photosynthesis), cohabitation (organisms, insects, and other animal inhabitants), and attachment (to the Earth). If we draw a tree according to these guidelines, it will not be miniature, nor a legend, nor a symbol, nor an object, but a point of life: a singular way of unfolding things as they are, in space, with the world around one. Through its characteristics, the tree allows us to envision a new space of reference. Structurally speaking, it has an exploratory capacity that allows it to develop in all directions of space and also to integrate the fourth dimension: time. Memory in motion, it captures, draws, and encodes data about the ground and climate conditions. It is an environmental marker and a palimpsest being. The tree, in an inverse shedding motion (the living covers the dead), integrates its (organic) memory just as it keeps track of its displacement. As an ecosystem, it combines scales through reiteration and fractal development.

Space, time, scale: trees offer our model an alternative system of reference, implying a new spatiality (fig. 11). As a frame of reference, the tree generates another relationship to the world, between anchorage in the territory and immersion in "the sea of the world" (operating on the logic of the milieu). The tree does this because it is attached to an extended nervous and fibrous network, and its health depends on the qualities of its relationship to its ground, its substrate, but also to the other organisms that surround it, including humans.

Just like a tree that, by receiving and emitting chemical information (in particular by exchanging carbon and oxygen), to a certain extent is "global," today's terrestrial inhabitants are involved in the world in a global way (through food, travel, materials, and so on). When we try to understand the global spatially, it implies a succession of intermediary places that are possible to locate. However, it is a more complex relationship than a linear succession of places. This is why we visually represent the point of life with a system of nested envelopes in which the territory is a series of threshold crossings, starting from the skin, the area of contact, and ranging to its most distant extension (see model pages). Like Hooke, when we look at the point of life under a magnifying glass, enlarging it until a world emerges, we open up its multilocalities. We progressively allow externalities to enter, which define the extent of our territory from the inside out. In a movement that is symmetrical to that of the first map, the world is enfolded within each perceiving entity. This map develops the internality of the point of life, which includes all of the frontier territories that constitute us, the first of which is the skin (whether we are human, plant, or animal). Thus, for a human being: skin (contact)—habitat (surrounding space)—parcel of land (vicinity)—municipality (community)—larger area (urban sprawl)—catchment area (broader geography)—region (cultural specificities)—nation (broader cultural specificities)—geographic area (international specificities)—continent—world. And for a plant: leaf (solar reception)—soil (material for roots to grow in)—local atmosphere (transfers with its vicinity)—world (cosmic exchanges). Thus, the point of life is not an organism that is closed in on itself, internalist, or "psychologized." On the contrary, it contains the world, and its interpretation of the world corresponds to the nestings in which it is included and which include it. As a consequence, the interior world is equivalent to the world itself, which we habitually think of as something exterior. In this hypothesis, there is not an imaginary world inside of us; rather, each being's territory is contained within it according to a complex system of membership made up of envelopes and trajectories.[28] This point of life isn't an imaginary space; it is a territory, the territory of a being with attachments and singularities. It is the living territory that every individual has to sketch for themselves.

From the skin to the territory-world, the map features a series of sections that establish direct and transversal links between the biological physiology and the physiology of the territory. From the viewpoint of method, this map is like the result of a game where you have to move without losing contact with the ground or with others. This is why there are no longer blank spaces but rather a continuum of elements (see map pages). The territories that compose us, that is, which are interlinked in our point of life, have as many effects on our bodies as our bodies have on them. This is why, in the continuum of surfaces, the ground is as sensitive as skin, and all of the phenomena we feel (climates, storms, heat) are also felt by the ground. This skin-map attempts to capture the way in which surfaces are linked in

---

28 The point of life is defined, not through itself, but through the network of places it contains, in the experiment with envelopes as presented here, or the places that it reaches, in the second attempt with trajectories that we will make in the following chapter.

a continuous sensitivity by the elements that cross the different strata, span thresholds, and multiply the leaps of scale. Placed next to each other, these surfaces form a meteorological tapestry, an epidermis bristling with spores, evocative of what we are "hung with" on the inside, to borrow a phrase from Deleuze.[29] The effects of continuity between meteorology and epidermic reactions become apparent: wind—goosebumps; rain—sweat; earthquakes—the bristling of hair. This network of dependencies constitutes, more than a morphing of the skin into ground or the nervous system into rivers, rather a plea for the recognition of the territory integrated into and for/by each living thing in interconnection with others, generating living landscapes whether they know it or not. Of course, the representation of the point of life as a map remains a rough outline. The representation must be read not as a closed organism but rather as a monad, a spongy point, an onion with multiple layers, a porous cell capable of traveling across scales (fig. 12).

The visual representation of boomerang effects (as seen in the previous model) caused by climate disturbances prompted us to create this model in which the point of life is involved in a series of climato-territorial envelopes whose traditional connections have been reversed: near (center)—far (periphery); global (periphery)—local (center). Thus we see that to be indifferent to the fate of the ground or the plants with which we share space is to be indifferent to the fate of our own skin. The point of life illustrates which envelopes we are sensitive to and which keep us alive. The violence of climate disruption on either also affects us. More than just sharing space, it's nearly a common identity, shared DNA, a physical connection. We must "stick together" to survive. Furthermore, the model begs the question of the connection between the ground and movement, between the

29 Gilles Deleuze, *The Fold*, trans. Tom Conley (London: The Athlone Press, 1993), 29.

ground and life. This is the subject of the diagram about bonding points (fig. 13): exploring the different combinations of the ground and the point of life by describing how the latter manages to "attach" to the former. The chosen perspective allows us to observe different kinds of attachment and assemblages of points of life. Certain points of life merge with the ground, while others have a fragile mooring. Someone very mobile and unattached may have a very large reach thanks to their online network; conversely, a deeply anchored being may be strongly moored but have a reduced reach. Moreover, anchorage and reach are not necessarily correlated.

The point of life *terraforms* the globe's localities, but it doesn't do so alone. It is surrounded by other points of life. We can interpret the Gaia hypothesis in this way: it doesn't mean that the Earth is a living thing, but that the Earth's ecosystem is inextricably interwoven with living things; that there is no physical milieu, because a living thing's milieu is other living things and is the product of their metabolic activity in the past and present.[30] How should we visualize their interactions? What happens at the intersection of their trajectories? We must now delve into these tangled lines.

---

30 Here we have adapted the Gaia hypothesis, developed by Lynn Margulis and James Lovelock, and formulated thus by Tim Lenton: "life and its non-living environment on earth form a self-regulating system that maintains the earth's climate and the composition of the atmosphere in a habitable state." Tim Lenton, *Earth System Science: A Very Short Introduction* (Oxford: Oxford University Press, 2016), 4–5.

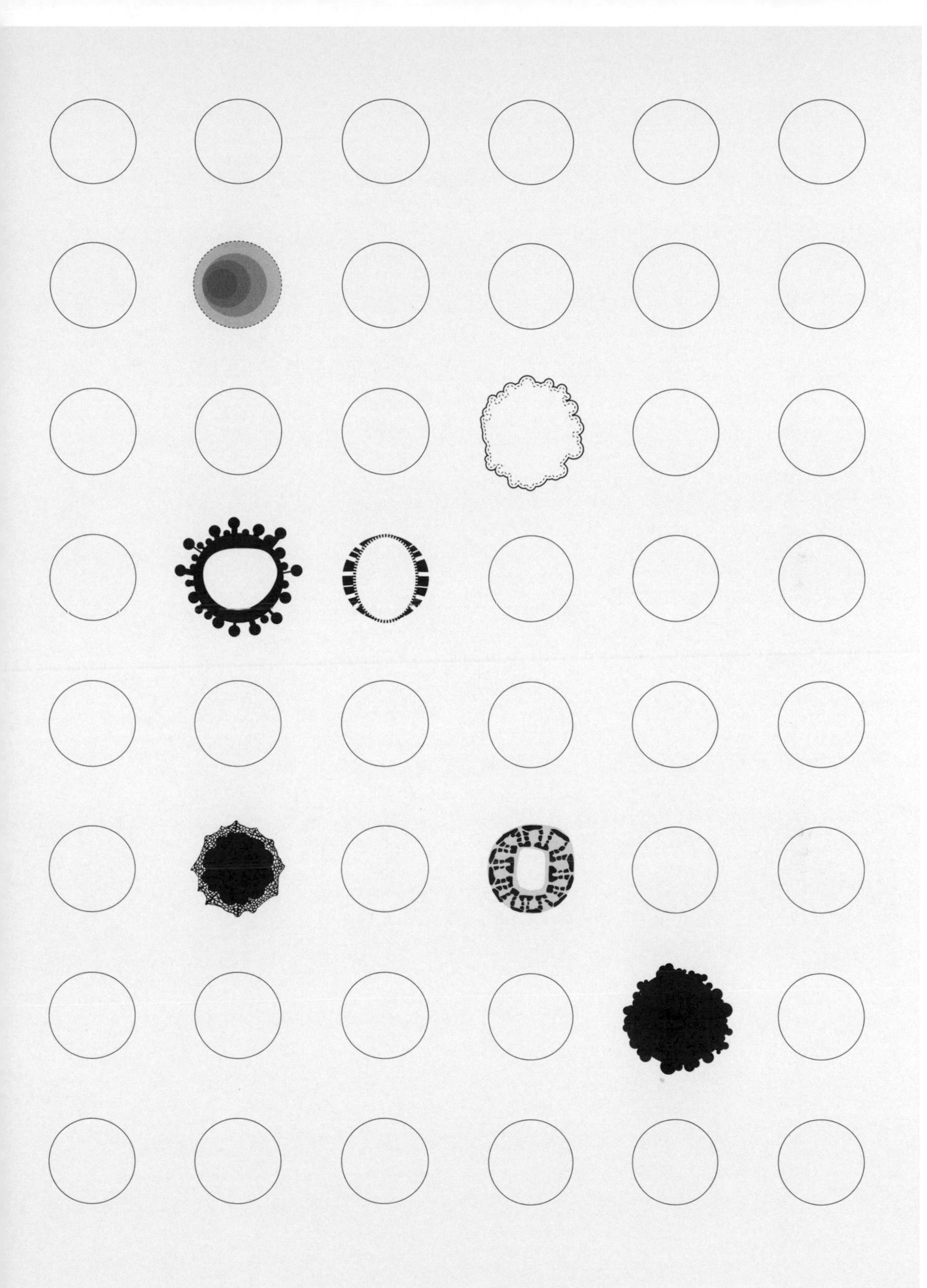

ig. 12 Variability of points of life

These points of life take on diverse forms and modes of bonding to their ground, and they redesign their territory in different ways.

CROSS SECTION

VIEW FROM ABOVE

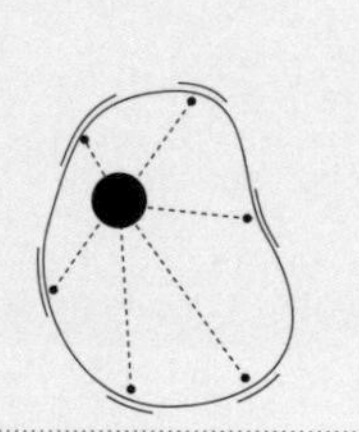

Anchoring
Territorialized cooperation
Mooring, attachment, commitment

Parasite
Ground as a host organism
Squats, forced occupation, dependency

Welding
Fusion of the body and the ground
Modification of DNA, absorption

Pressure
Hyperlocalization
System of fixation, active bonding

fig. 13 How points of life bond to the ground

What forces of attraction attach us to the Earth? What relationship to the ground does each point of life create?

Points of life: modes of bonding
Each figure is named, described, and illustrated with examples.

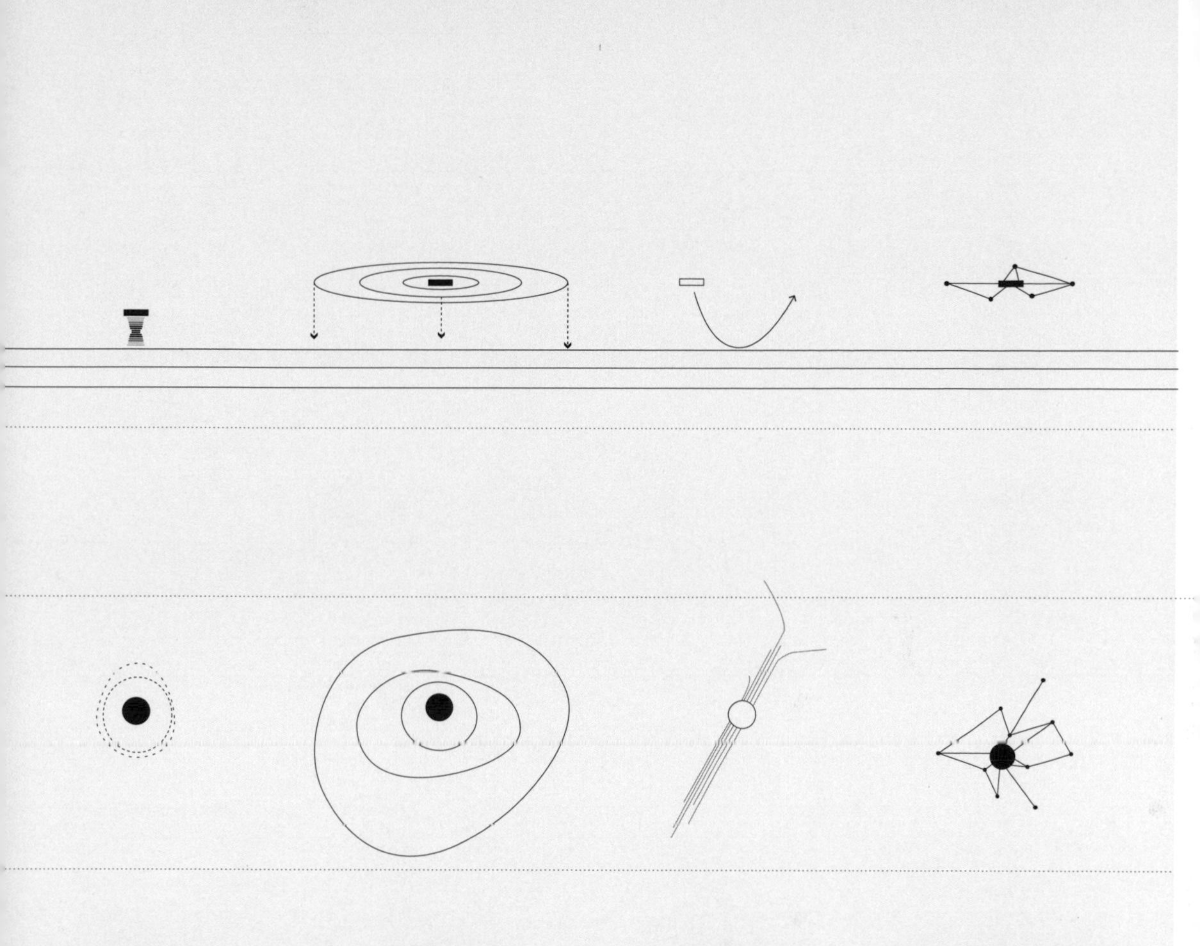

Magnet
The ground as superconductor
Equivalent magnetic poles, mutual attraction

Radiating
Large reach allowing weak attachment
Expanding web, diffusion wave

Ricochet
Lack of bonding or inability to do so
Increase of friction, sliding

Network without contact
Virtual relationship allowing relative independence
Parallel reality, mirrored world, intangible territory

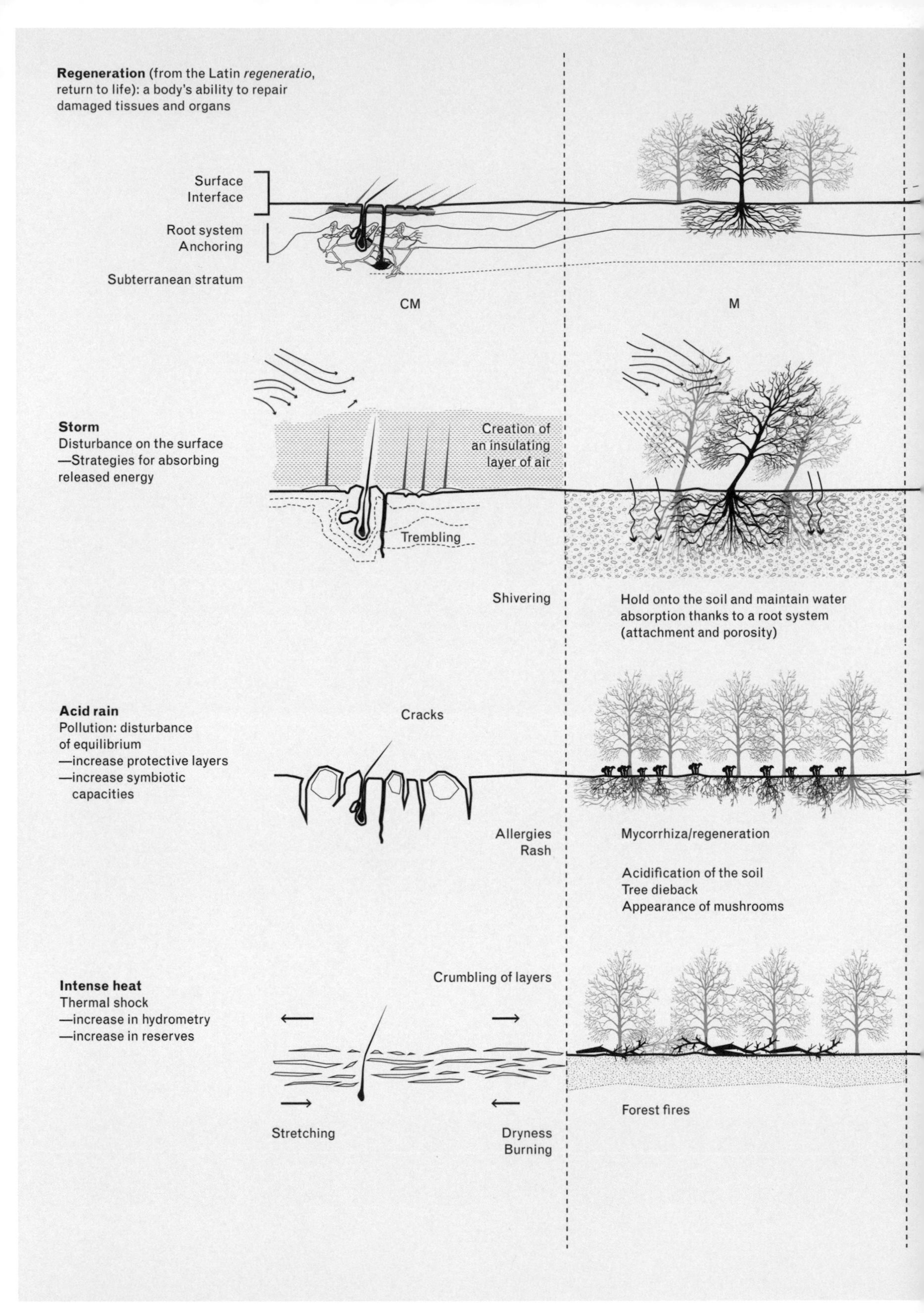

fig. 14 From the skin to the ground—the consequences of phenomena linked to climate change

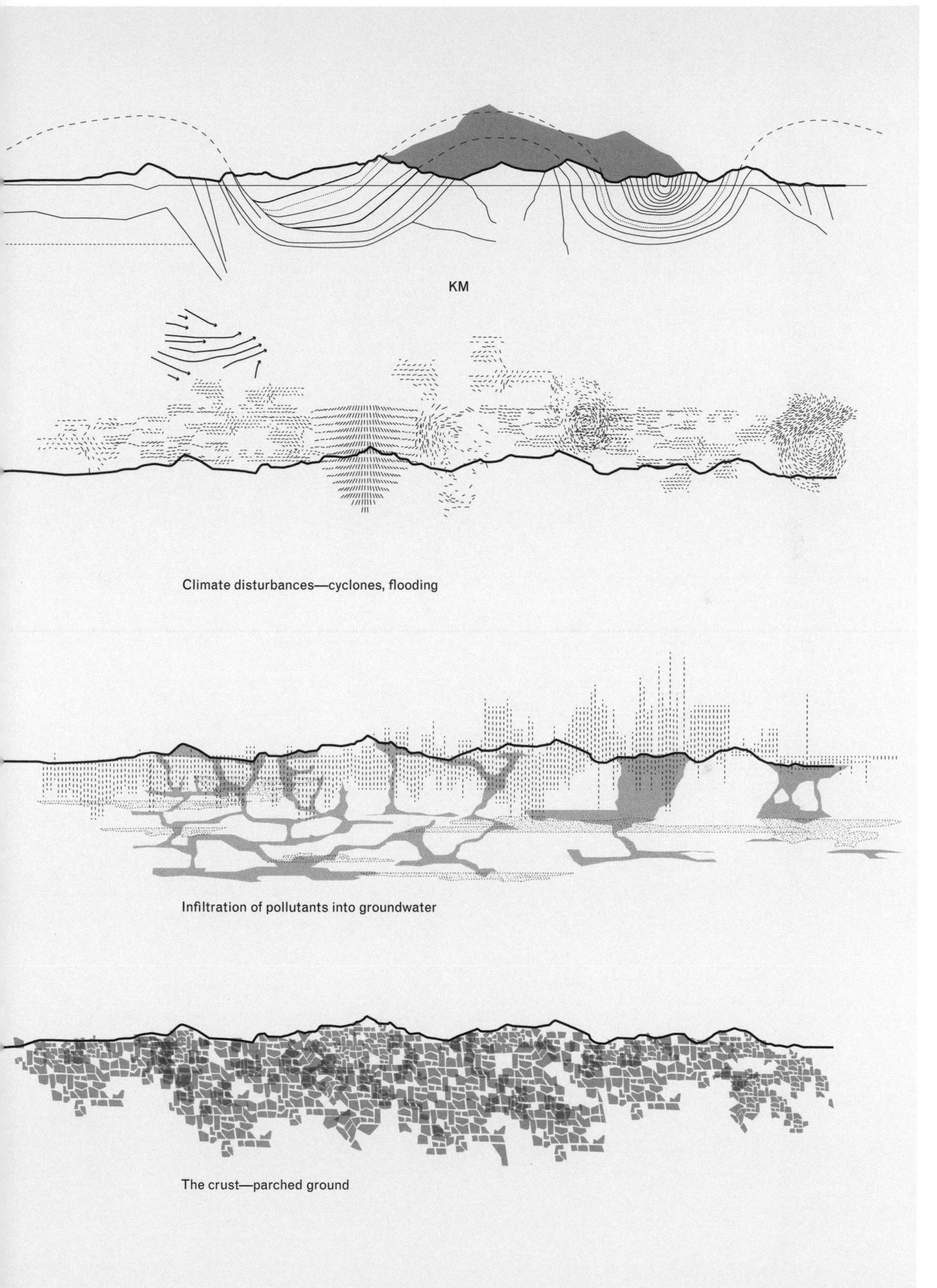

Climate disturbances—cyclones, flooding

Infiltration of pollutants into groundwater

The crust—parched ground

MODEL II
POINT OF LIFE

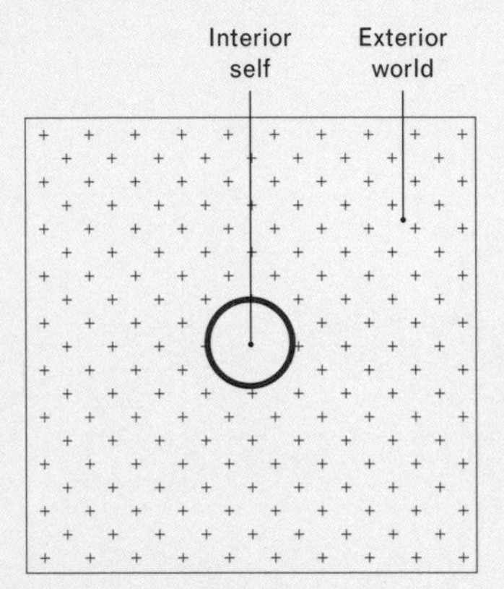

Modern conception of the relationship

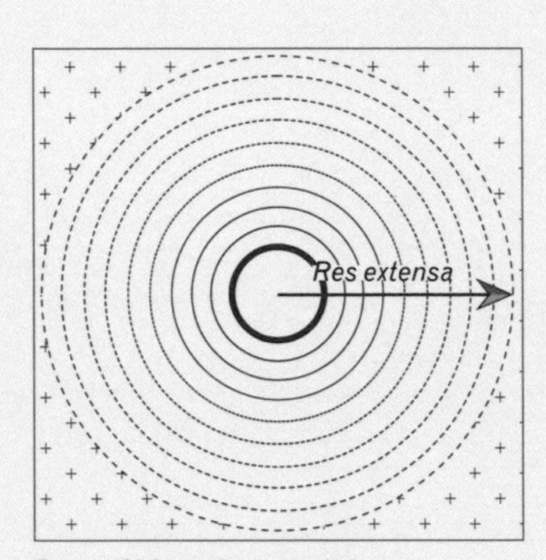

Point of life without qualities

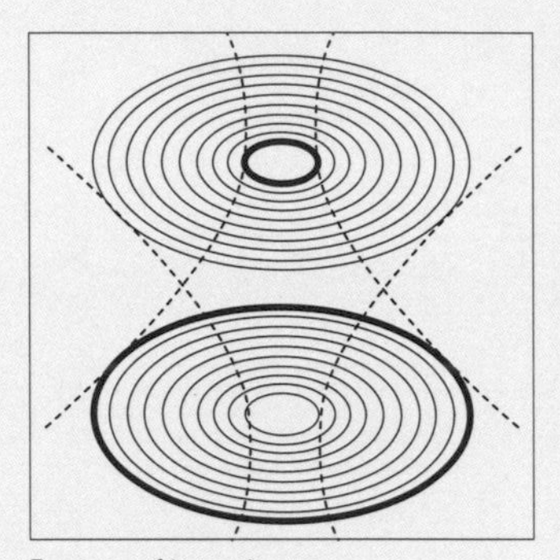

Process of inversion

1

Diagram of the traditional (modern, Western) relationship between the self and the world: a hermetic interior and an open and infinite exterior. The organism occupies a space in an environment that is physically separate and indifferent to its properties.

2

Map the progressive extensions of the self toward the exterior. Identify the process for constructing space in its traditional conception (the Cartesian *res extensa*), that is, a space defined solely by its dimensions. The outside (what is outside of the point of life) can then be annexed, since it doesn't belong to anyone until someone takes possession of it.

3

Invert the initial positioning: what happens if we admit that the world is already contained within the point of life? That the skin is our final limit, but even so, it's not the seat of a unique self, but rather the medium through which we experience, discover, and shape our habitats, territories, and the world, in relational intensity and not in an antirelational annexation?

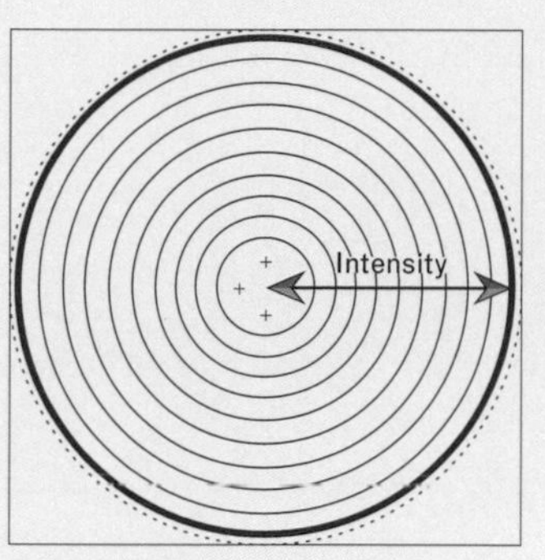

Point of life, qualities determined by the world

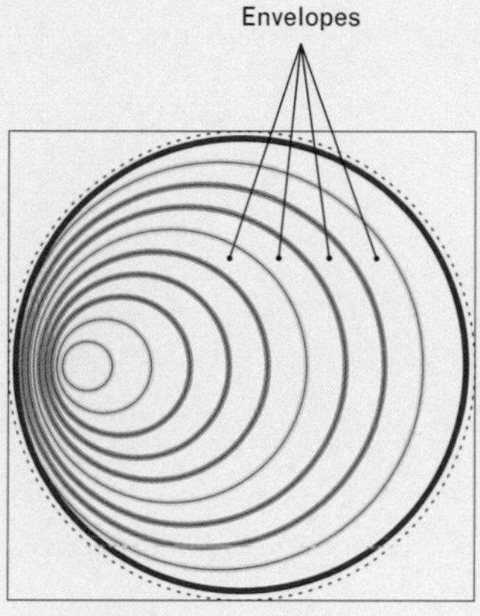

Discovery of the living territory

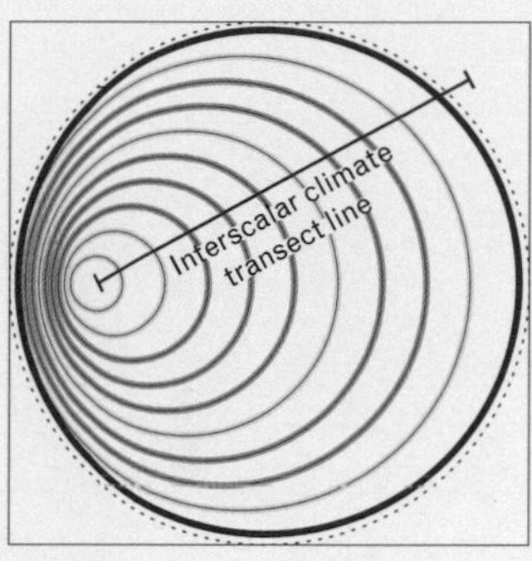

Climates

4

Assemble the point of life, whose qualities are determined by the world and whose reach is determined by its intensity rather than its extension. The point of life is not hollow or hermetic; it accommodates the spaces that construct it just as it contributes to constructing them in return.

5

Construct successive nested envelopes inside the point of life (the outside within). Thus, the living territory is constructed for each being according to the qualities of its envelopes: from skin to geographical or bureaucratic borders. Some envelopes are physical (skin/barriers), while others are invisible (bureaucratic/climate). The model seeks to describe the world located in the point of life: what are its envelopes? How many are there? How do you arrange them? Which seem closest to you, and which seem more distant?

6

Observe the variation of the envelopes due to climate change by means of what we call interscalar climate transect lines. These transect lines cut across the envelopes of a living territory (fig. 14). They describe the effects of climate disturbances on the various levels they cross. In fact, skin, trees, soil, borders, and buildings are all sensitive and vulnerable to the combined effects of pollution and extreme climate change. These tangible changes affect the land and can also be read in its "physiological parchment." There are no longer global or local phenomena that can be isolated or separated—they are all connected by the climate.

MODEL II
POINT OF LIFE

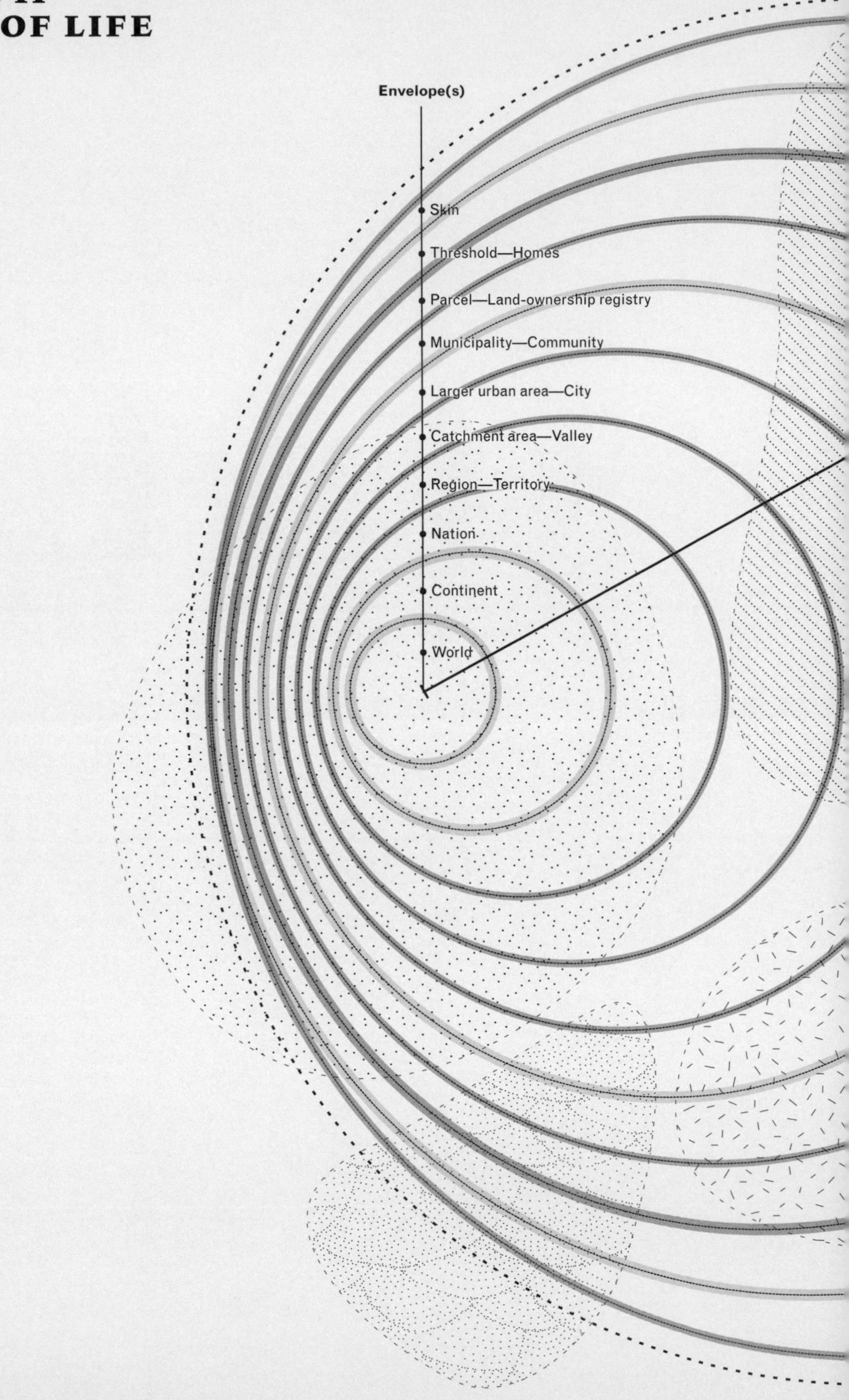

Interscalar climate transect line
Climates

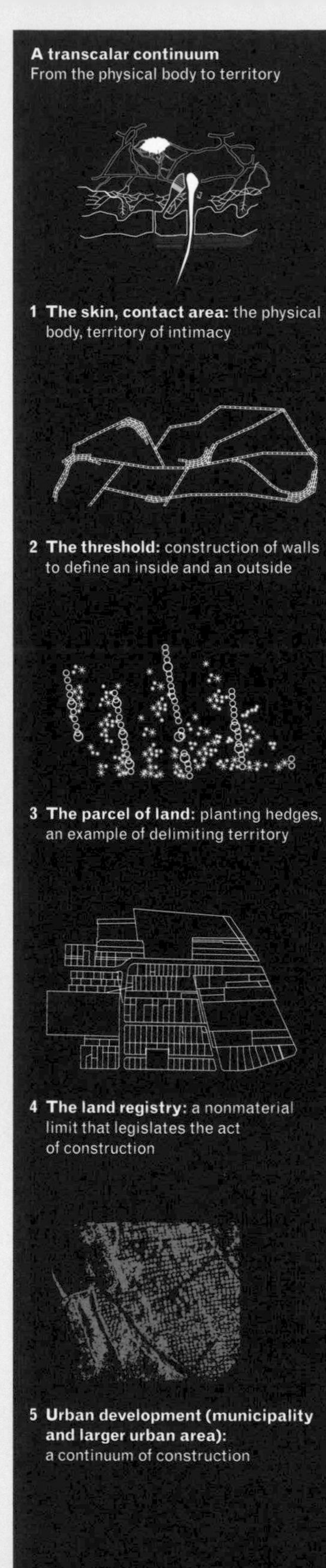
A transcalar continuum
From the physical body to territory
1 The skin, contact area: the physical body, territory of intimacy
2 The threshold: construction of walls to define an inside and an outside
3 The parcel of land: planting hedges, an example of delimiting territory
4 The land registry: a nonmaterial limit that legislates the act of construction
5 Urban development (municipality and larger urban area): a continuum of construction

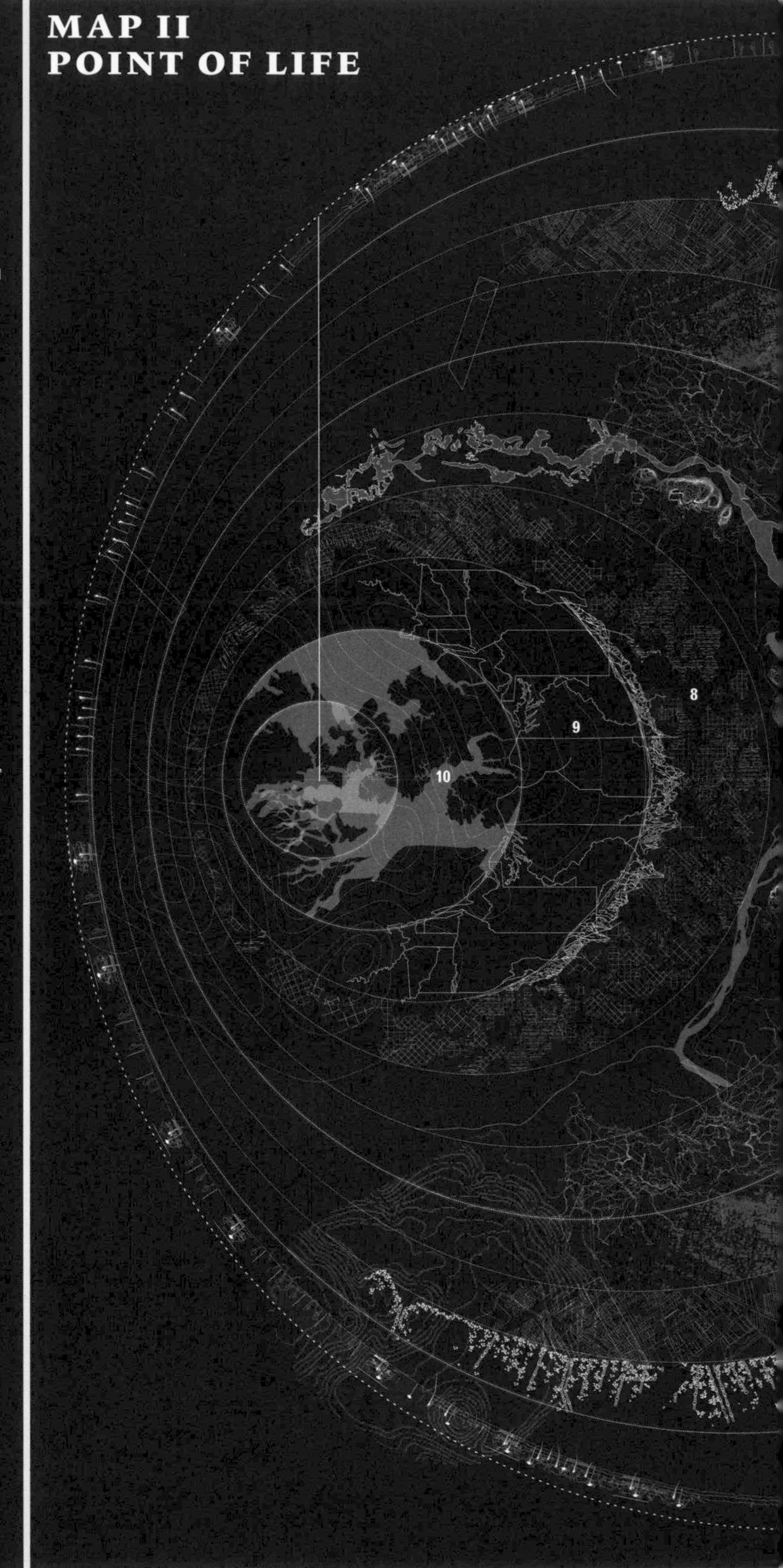
MAP II
POINT OF LIFE
8
9
10

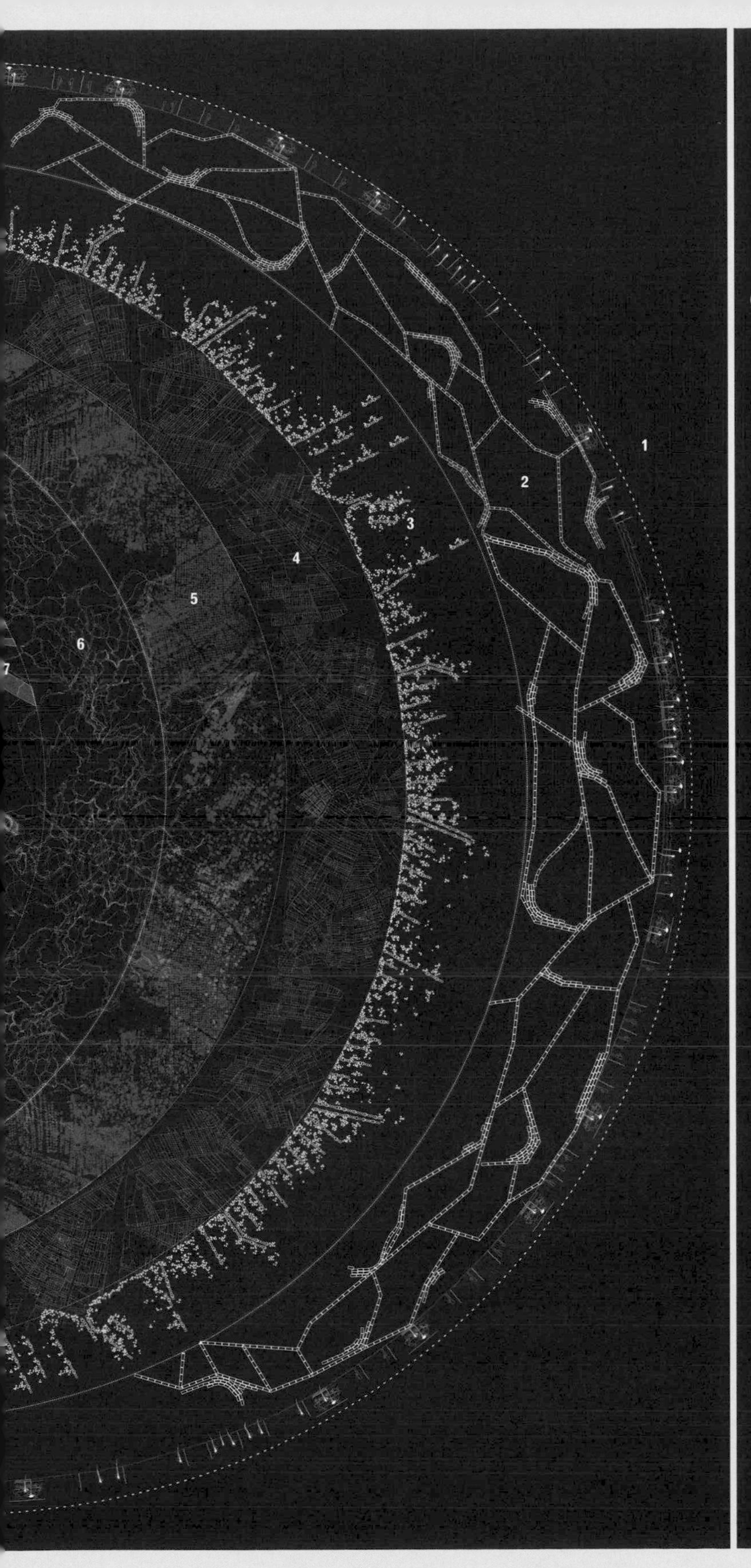

6 Forested areas: surfaces of exchange and filtration, ecological corridors

7 The catchment area: a hydrological unit

8 The bio-region: an ecosystem-based unit

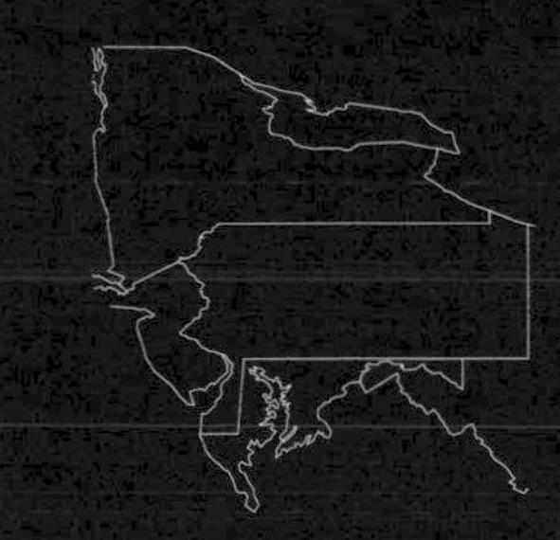

9 The nation: immaterial borders

10 The continent: divides land from ocean

TRANSFORMATION II
POINT OF LIFE

The inclusion of scales . . .

. . .within the encompassing body.

fig. 15 French National Geographical Institute map of the southeast Grand Paris area

MODEL III

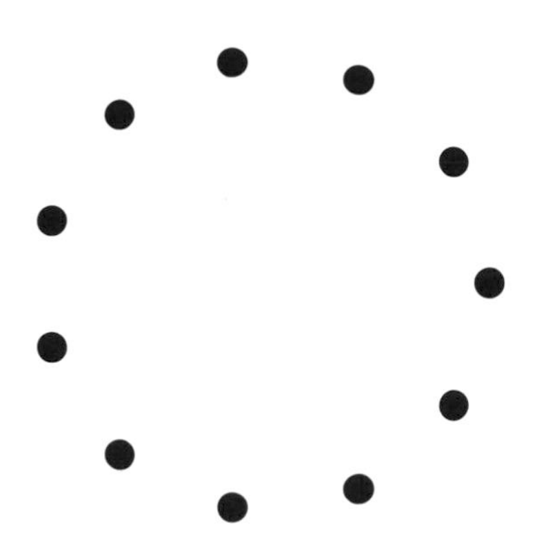

LIVING LANDSCAPES

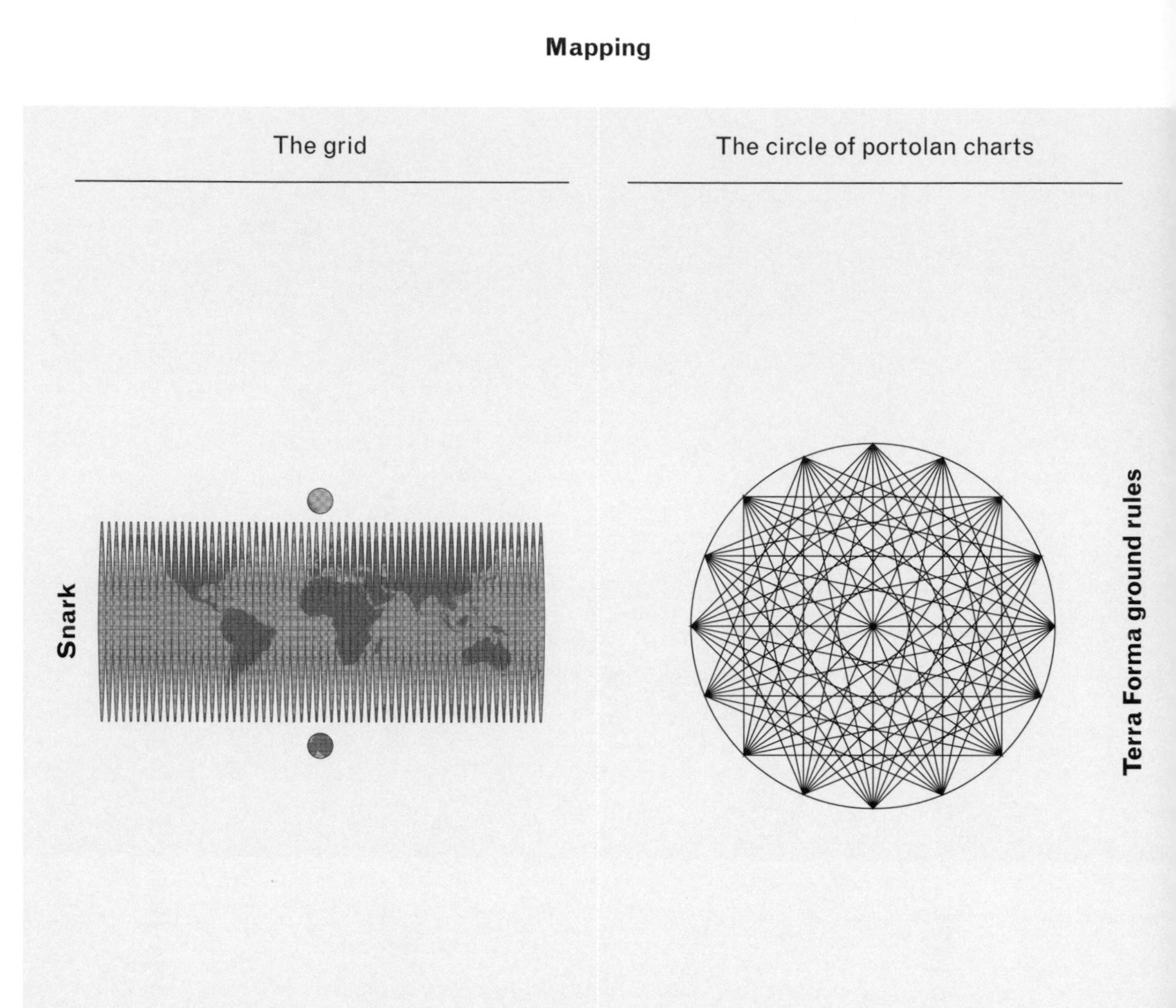

Former framework

New framework

Willian Rankin, "The Universal Transverse Mercator Grid System," in *After the Map: Cartography, Navigation, and the Transformation of Territory in the Twentieth Century* (Chicago: University of Chicago Press, 2016).

LIVING LANDSCAPES

In Australian Aboriginal cosmology, mythical beings took journeys that left traces, furrows, imprints, holes, and so on. Their actions shaped the landscape. By tracing these myths and applying them to our everyday life, we will attempt to move from a conception of a lifeless landscape to a living landscape.[31] In fact, it seems pointless to resort to GPS (or only to GPS) to observe living things in the process of *terraforming* their territories, because GPS assumes a stable world to it which applies a globe-shaped grid with fixed coordinates based on horizontal and vertical positions (latitude and longitude, respectively), recorded by satellites in space orbiting the Earth (fig. 15). (It seems that a heavenly point of view really does exist, thanks to technology.) But this tool offers only an aerial visualization of each source, ignoring its relationship to the ground and its motivations. Our inquiry focuses on the motivations of the point of life as it proceeds on its journey, in order to visually represent its "dwelling place,"[32] and subsequently to allow it to orient itself, not in Euclidian space (the space of GPS), but in an environment composed of other points of life mapping out their own dwelling places. "Dwelling places" here refers to each living thing's relationship to the ground, its way of making a mark on the land scape with its movements or its way of being in the world. To retrace each trajectory and each geometry is to try to perceive

---

31 As Philippe Descola explains, the entire notion of landscape is transformed through this process: "The goal here is absolutely not to represent a portion of a country viewed from a fixed point, which is the usual way of representing landscape, but rather to represent paths of morphogenesis that may be connected without ever being integrated into a homogeneous space." Philippe Descola, "Figures des relations entre humains et non-humains"/"Ontologie des images" (Class lecture, Anthropologie de la Nature, Collège de France, Paris, 2000–2017).

32 Latour, *Down to Earth.*

the non-anatomical body of living things, the form of being that they display. Together, these forms of life make up a landscape.

To distinguish the contours of this landscape, there is no need for a topographical survey or a background map; you have to make an inquiry, which is to say, patiently follow these beings in their movements. This undertaking requires the tool of walking, taking a journey as a way of grasping the world (sometimes quite literally), and physical contact with the ground as a characteristic of living things.[33] Starting from the home base (the house, den, hole, etc.) of each entity under consideration (whether human or nonhuman), we follow its route, recording the places it visits and the connections it establishes. In doing so, we abandon topographical coordinates, the cartographer's and surveyor's tool par excellence, to make room for the body, for tracking, and for searching.[34] This is why we designed this model like a portolan chart—a maritime map invented in the thirteenth century in which the route determines the map. Constructed without a system of projection and without a terrestrial point of reference, portolan charts offer different landmarks: they indicate home ports and currents and provide a mesh whose lines correspond to the compass rose. In the same way, our portolan chart shows the movements of animate beings: their point of attachment, their trajectory, the area they canvas. The portolan chart functions as a navigation map allowing orientation in a common landscape made up of overlapping dwelling places. It is a tool for navigating on land, among living things.

The model is constructed by tracing a living thing's reach, from its point of origin which constitutes the departure from a preferred area (home and access to a main resource) and progressing toward different areas of action linked to the community

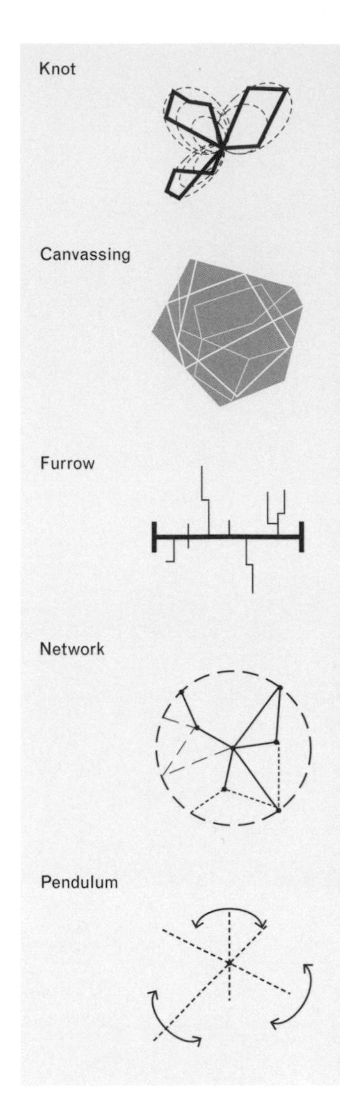

fig. 16 Geometric grammar of dwelling places

33 Careri, *Walkscapes.*
34 Morizot, *Les Diplomates.*

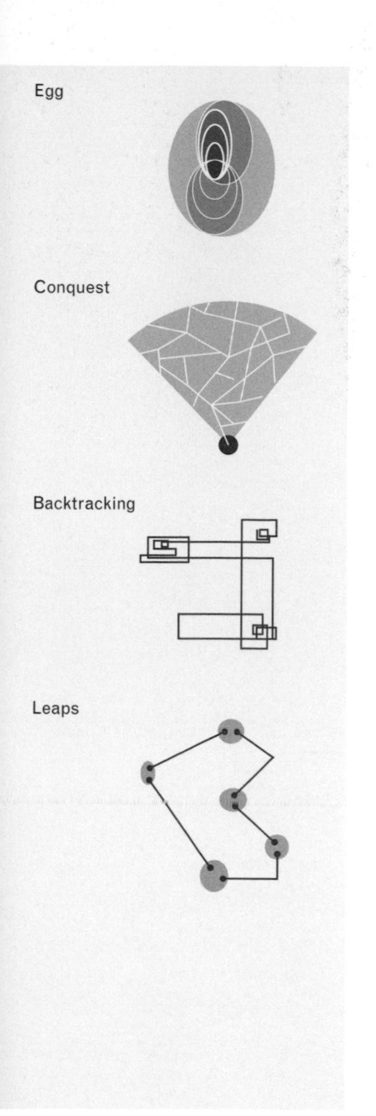

(families, friends, packs, herds), leisure activities, or desires for expansion (see model pages). The dominant reach characteristic of the being under observation can then be identified. This physical or immaterial reach is defined by repetitiveness of action, by duration, or by the fact that it determines this being's subsistence. By drawing the trajectory or trajectories of each living thing in its urban and/or suburban space, its canvassing areas, and its networks, we see its "moving grasp" in shared space. The next step is to position the different human and nonhuman points of life around a shared matrix on which their points of attachment are recorded—home ports essential to navigation.

With the notion of a living landscape, we are trying to discover a physical relationship to the territory, constituted by movements, shifts, and habits—by the body and gestures rather than by the all-powerful eye.[35] This approach challenges a certain way of seeing landscape (gazing at part of a country) and a certain conception of space as a homogeneous and continuous expanse (the *res extensa* of Cartesian physics). We thus pass from the point of view to the area of action, from the eye to the body, from the gaze to the gesture, from the fixed point to the route, so that landscape becomes the constantly updated result of the actions of living things rather than background. The map reveals forms that we call "habitual territories," which represent different ways of habitation. These territories appear here in the diversity of their geometry as so many styles and paces (fig. 16). By taking note of these emerging forms and classifying them, we can reconstitute a sort of transversal grammar for different animate beings, allowing us to identify the spatial imprint of their uses of the territory. This leads not to a stable and unified

35 With regard to land art, Careri writes that certain artists "have transformed geographic space into cartographic space, a surface to write on," and that they use their bodies "like an instrument of space and of time." For him, "the body in movement reflects the physical structure of the territory" (Careri, *Walkscapes*, 155).

geometry, but to a multitude of geometric figures: the figure of canvassing, signified by veins like those on a leaf, testifies to the exploitation of entire surfaces; the figure of the leap, which consists of going from one micro-territory to another by borrowed means, reflects a form of swarming; that of the groove develops along a major axis like a rod. The figure of the knot or the flower evolves in multiple nonconcentric loops around a point; that of the egg traces routes and detours within a minimal pocket area; that of conquest unfolds rhizomatically. Finally, though this is not an exhaustive list, the figure of the network offers an extensive system of immaterial connections and a minimal physical imprint, like a membrane. This inquiry thus allows us to transform invisible habits into a map of the territory, translating the geometries that emanate from points of life into notations and topographies.

At the center of the model, the crossing paths of points of life extend from the ports into the world-sea, constituting a common landscape. The points of life are motivated to venture into the middle of the portolan chart by the desire to extend their domain, encounter alterity, or to form a group or a population. This is the object of the final series of drawings, which focus on these "encounters." By following these sinuous lines to the common sea—the "windrose network" of the old portolan charts—we observe a new topography taking shape: movements sketch out shared micro-territories, while routes overlap, mingle, and coordinate, sketching a complex and rich territory. By superimposing human and nonhuman routes, new types of common spaces emerge that had been invisible until now (fig. 17). Instead of thinking of space in terms of circulation and the flow of people to be managed, the model suggests sketching other spaces where a new topology of public space can emerge.

The area chosen to test the model is Grand Paris, a preeminent site for major urban projects, thanks in particular to the establishment of the Grand Paris Express. Encompassing extremely urban areas and countryside, with density gradients characteristic of the transition between the two, this territory offers great diversity for studying the variation of figures by area of habitation (whether human or animal). The data required for plotting the drawings were collected from a sample of about ten inhabitants, as well as eco-ethological data. On the map, nearly 9,000 inhabitants share a living area of one square kilometer. But they sometimes travel well beyond this one-kilometer limit in their daily lives to meet their needs, and their routes overlap those of entities from other territories, which are quantified in a different ratio of population per square kilometer. Thus they increase the density in Paris, which is already very high. It is difficult to grasp the territory of a being in motion, unless we start from that motion in order to try to understand the territory. An office worker takes the subway daily to connect the different places in her life, which correspond to her many activities. The sketch of her living landscape develops with thick and articulated branches that open into malleable bubbles. A parasite jumps from body to body to survive. It follows a rhythm and a path designed to open windows of survival. What do these two living things have in common? They have the same figure of motion: the leap. This means that they make territory in the same way, but at different scales and by different means. Perhaps they've already crossed paths. The swift, a small seasonal migratory bird found in France in the summer, crisscrosses the air, patrolling it so as not to let any insect escape its area; a professional salesperson canvasses an area where he has been assigned to sell his product.

They both conquer an area on the fly. Do you recognize yourself in one of these figures? Do you share this figure with a being of a different species? Are you ready to change your perception of landscape?

The map thus highlights the gap between large-scale infrastructures and the fragility of our figures of movement, which are hidden and overwhelmed in the multitude. Above all, it reveals the diversity of these figures, no matter where they are or who makes them; in the end, we cannot establish typologies by place, habitat, or species. Rather, the map highlights affinities between species and between inhabitants of very different places, disrupting the usual categories of classification: city/countryside, human/nonhuman, suburban/urban, active/inactive (see map pages). While it is interesting to consider the similarities between figures of motion because it brings humans and nonhumans together, it might be equally fruitful to consider the influences of one figure on another. Are there inflections and bifurcations, or unforeseen effects when the figures are superimposed? The question calls to mind the wolves studied by Baptiste Morizot, which take the routes laid down by human infrastructure, its roads and paths: a furrow dug through the territory by another influences their figure of motion. While some figures cross each other without contact, others create friction, disturb each other, and may be blocked (fig. 18). Sketching these new kinds of maps means taking into account the superimposed and tangled lives that construct space, simultaneously envisioning different places and beings, modes and activities, metaphors and affects.

In short, the Point of Life and Living Landscapes models offer two ways of showing that space does not exist prior to living things, but rather is made up of the various living things that are present. The models offer two tools for the same question, forming two parts of a whole: one moves from the self toward the world (envelope), the other from the self toward the other (trajectory). Together, the two maps attempt to represent an extramental reality, a complex system of belonging made up of these envelopes and trajectories. If we consent to following them, several propositions arise from these models that might seem counterintuitive because they invert our usual relationship to space: space is the non-anatomical body of living things; the map is the territory because perception is a world, a territory; a map is never a map of space, but of animate bodies; every living being is both indigenous (producer of its dwelling place) and migrant (constantly crossing borders and occupying neighboring dwelling places). We are now wandering, unstable trajectories, constantly generating our space by inhabiting it. From this point of view, we have always already lost our own *Umwelt*, we are always in the midst of remaking it again, and the world itself is always slipping away from us. If the notion of belonging to a territory needs to be reexamined, how should we think politically about the composition and layout of our dwelling spaces, our habitats? We must now focus our attention on the old question of borders.

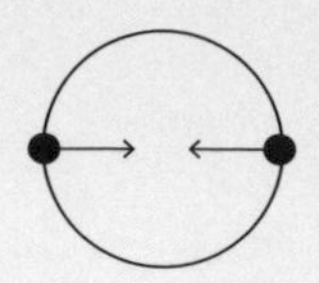

Overview of the creation of a landscape between a researcher and a childcare worker

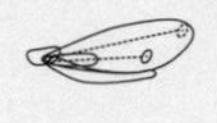

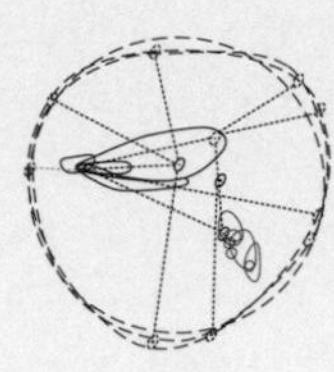

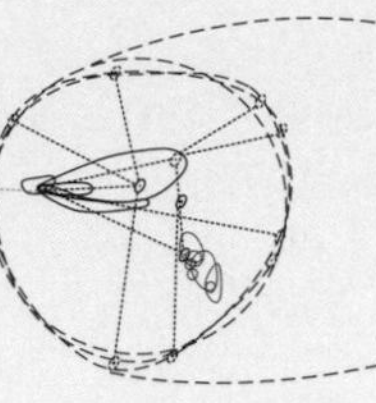

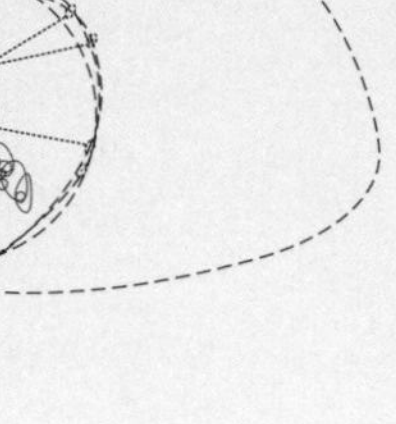

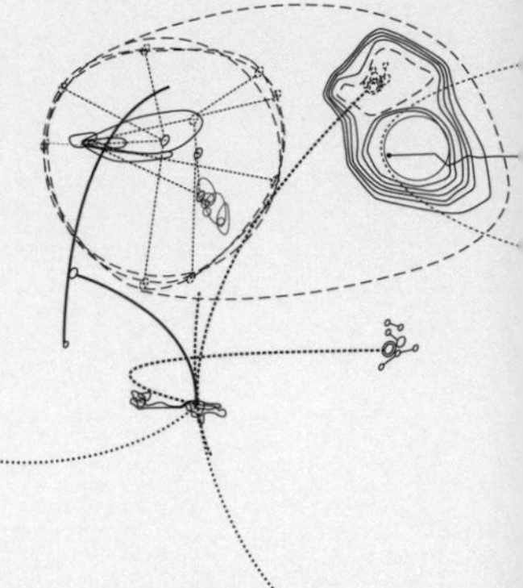

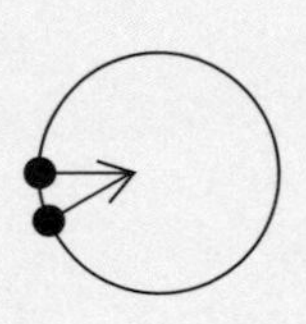

Overview of the creation of a landscape between an office worker and a teacher

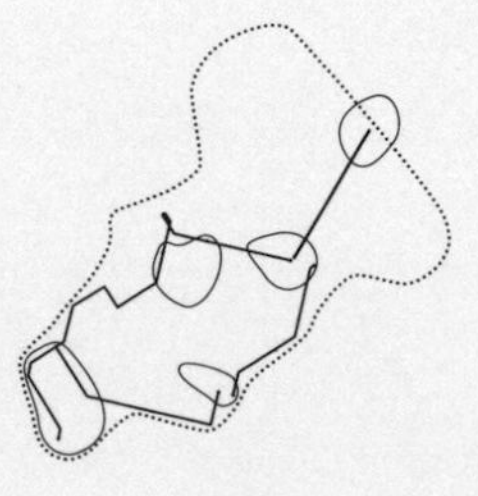

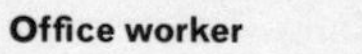

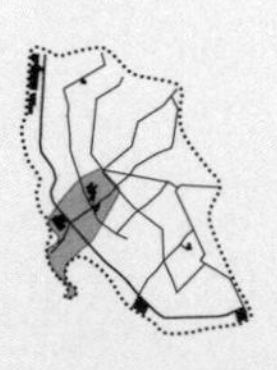

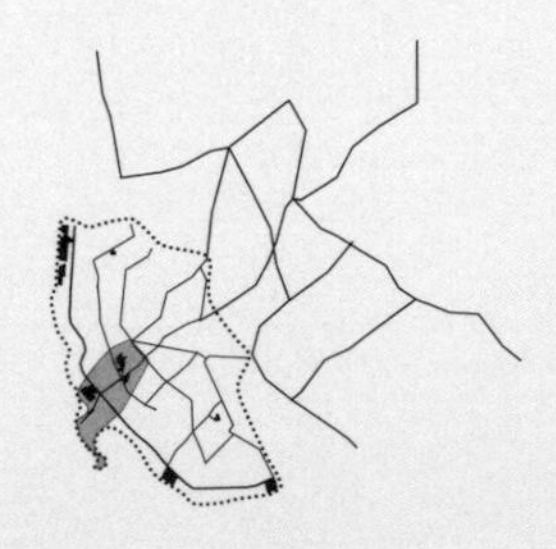

fig. 17 Overview: analysis of the assemblage of points of life and the formation of a common landscape

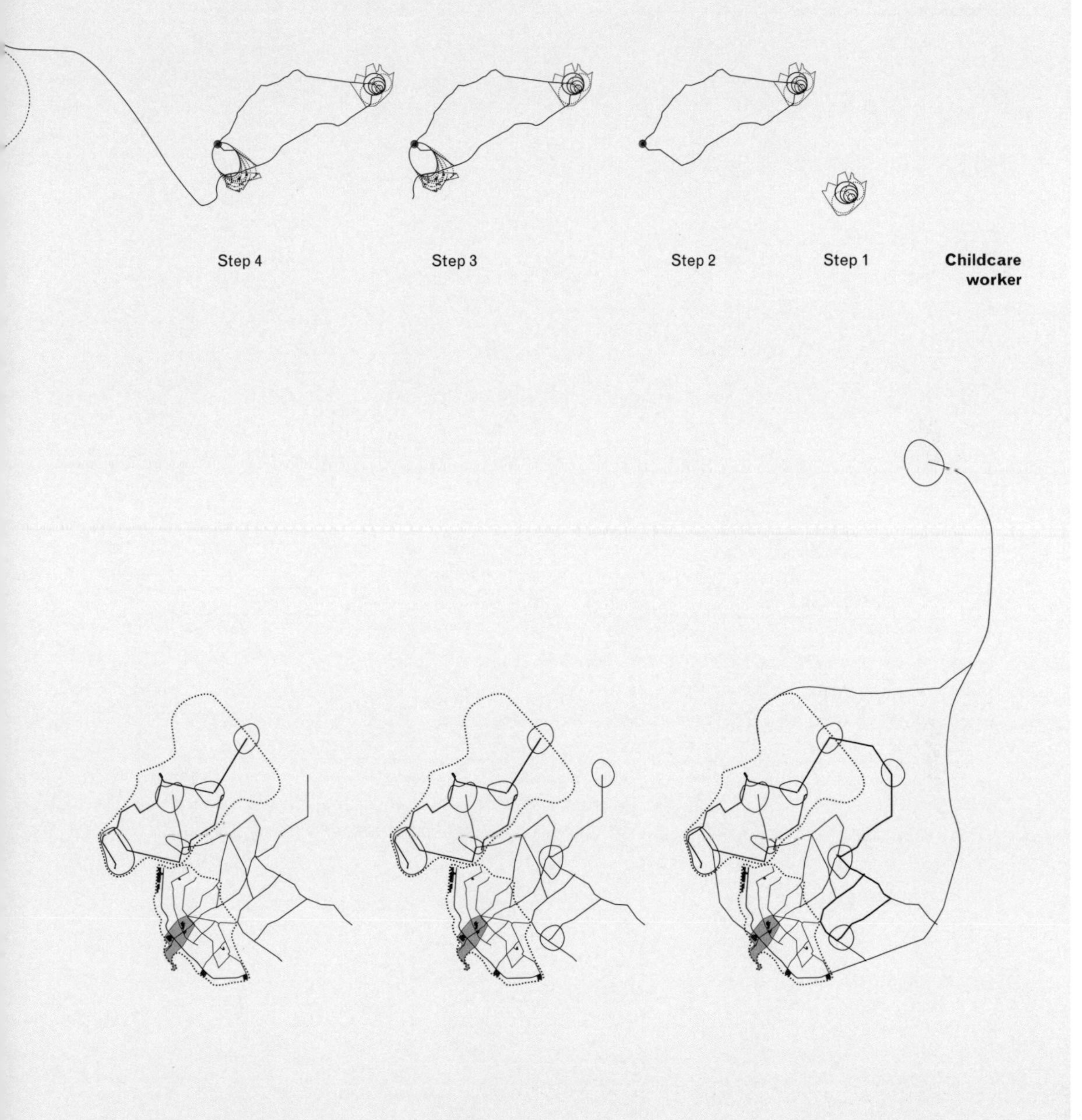
Step 4
Step 3
Step 2
Step 1
Childcare worker
Step 4
Step 5
Step 6

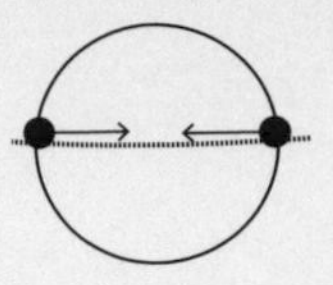

Overview of the creation of a landscape between a management-level employee and a stone marten

Management-level employee

Step 1

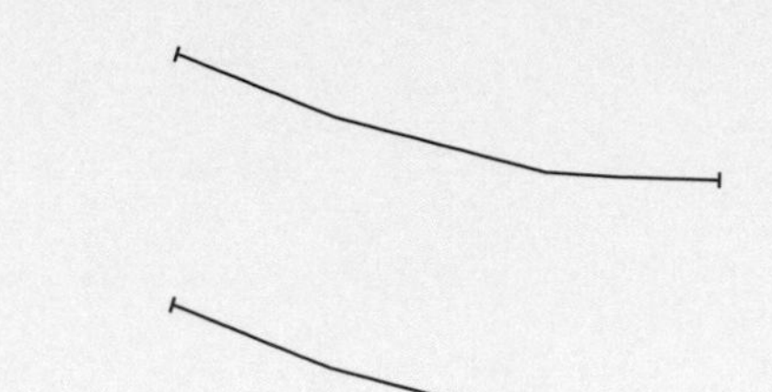

Step 2

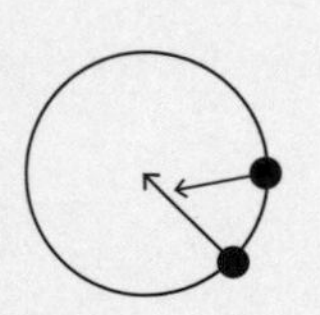

Overview of the creation of a landscape between a salesperson, a mole, and a microorganism

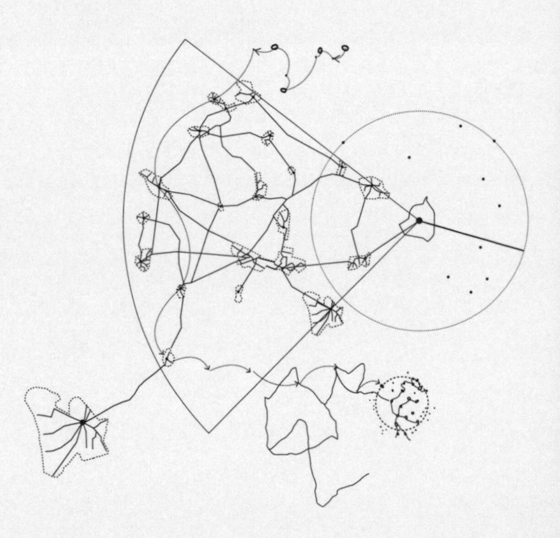

Step 5

fig. 17 Overview: analysis of the assemblage of points of life and the formation of a common landscape (continued)

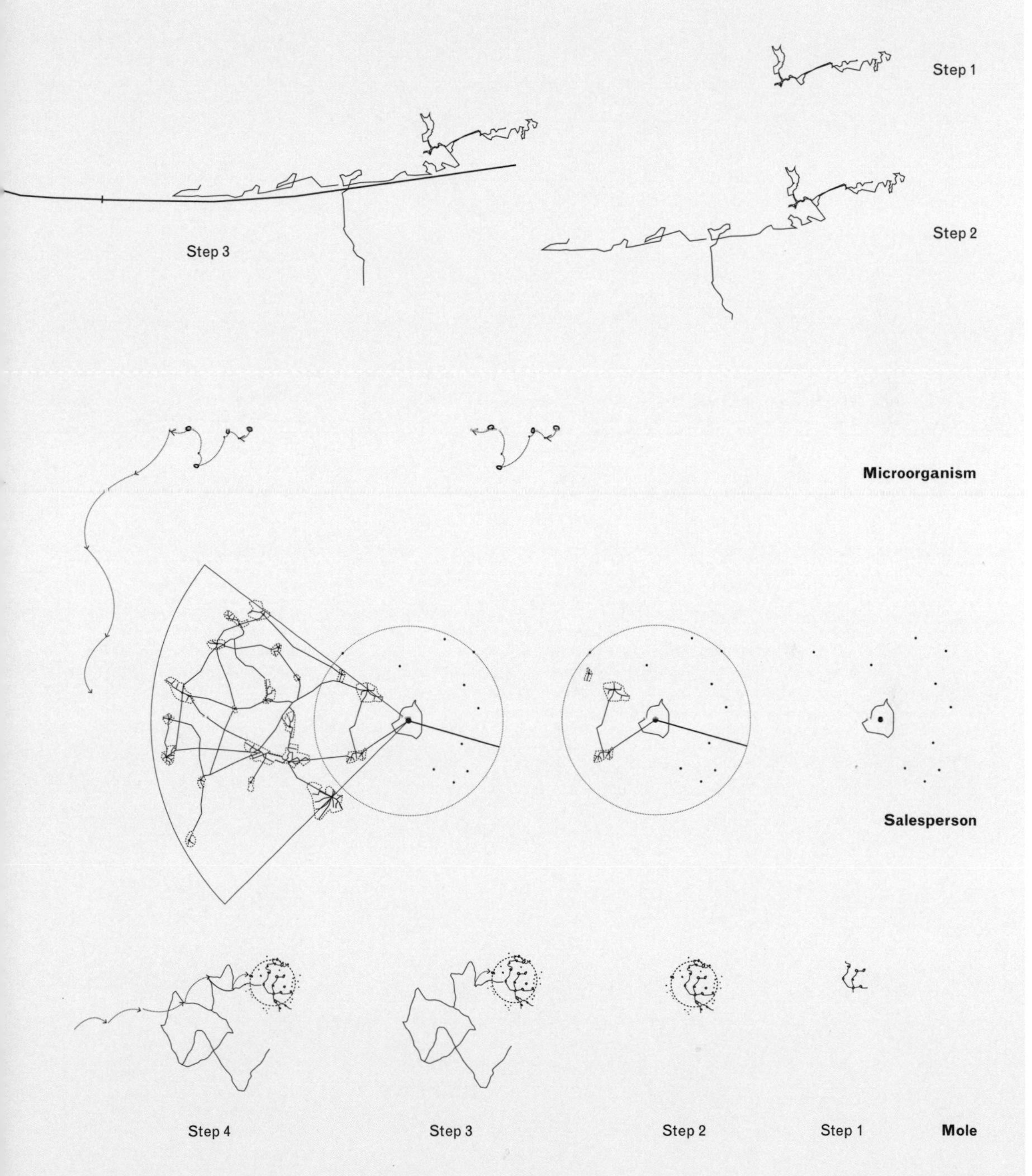
Stone marten
Step 1
Step 2
Step 3
Microorganism
Salesperson
Step 4
Step 3
Step 2
Step 1
Mole

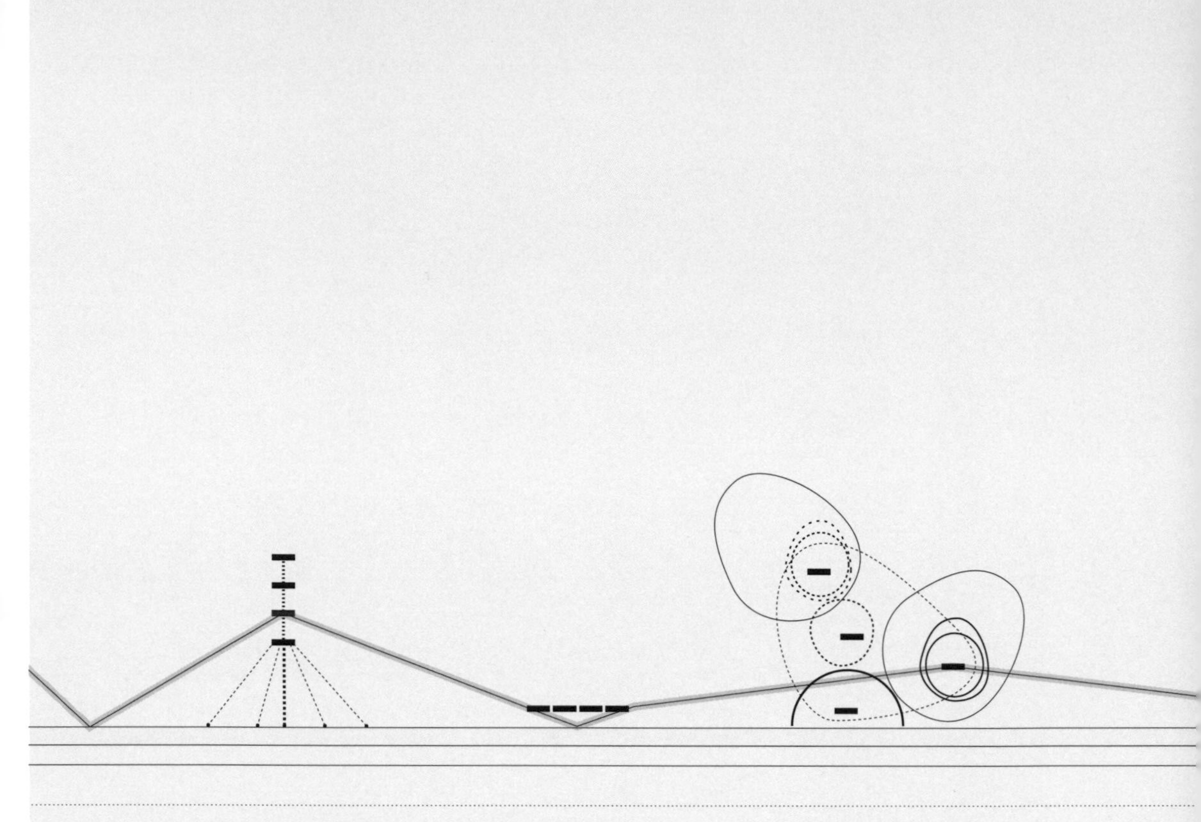

Layered landscapes
Cohabitation through layering

Platform landscape
Juxtaposition that allows exchanges between different points of life

Archipelago landscape
System of islet-refuges that points of life bond to opportunistically

fig. 18 Shared landscapes

How do points of life organize themselves to collectively compose shared landscapes?

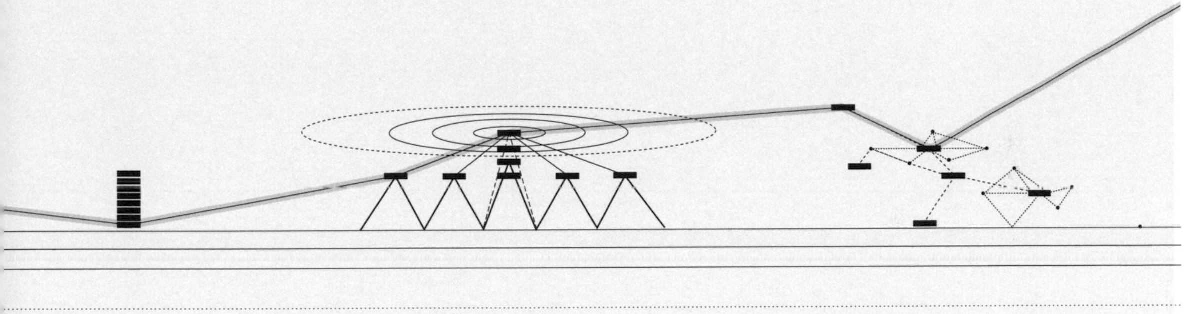

Agglomeration landscape
Fusion of points of life until a common body is formed

Skeletal landscape
Planned organization

Network landscape
Protean, evolving architecture

MODEL III
LIVING LANDSCAPES

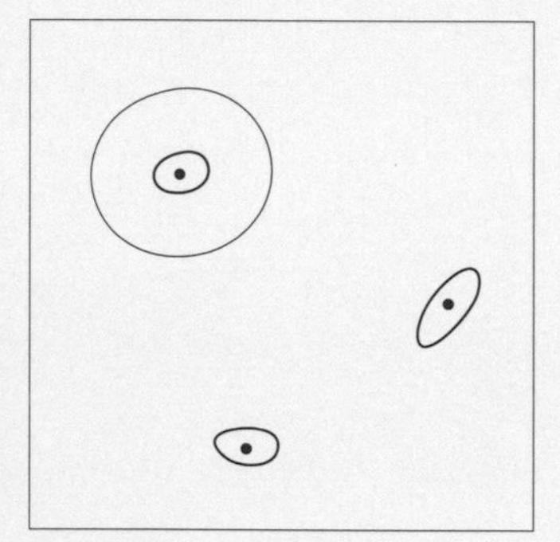

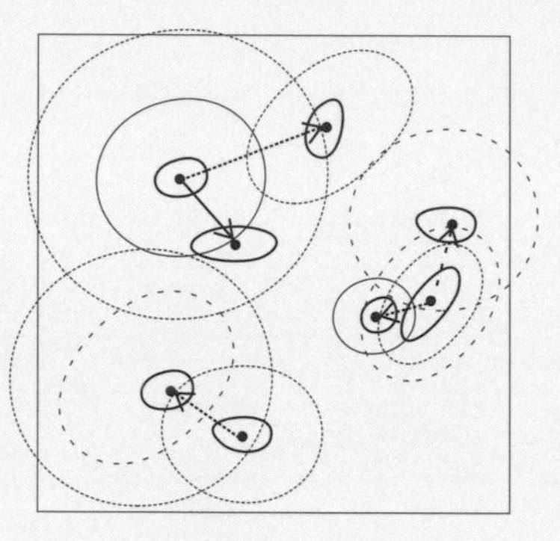

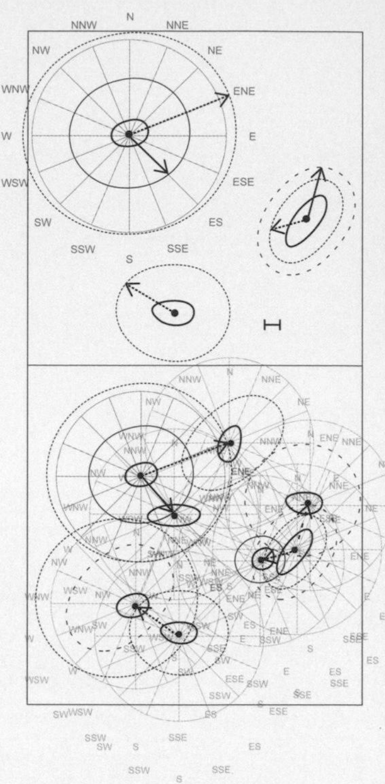

1

Make an inquiry: locate human or nonhuman points of life ("animate entities") and their home ports. A home port is a point of life's "preferred area" or "subsistence area."

2

Collect routes and record the trajectories of points of life: their journeys or movements. Trajectories give tangible form to "habitual territories" (these can be small or very localized in the case of plants), that is, the territories influenced by the recurring presence of one or more points of life. These extensions of points of life are figures in motion, tracings and traces that generate a geography. Narrate this life episode, this track followed, this landscape constructed—individually or collectively.

3

Layer the habitual territories. The common space constituted by this layering is given visual form by following movements, lines that are then drawn on the map.

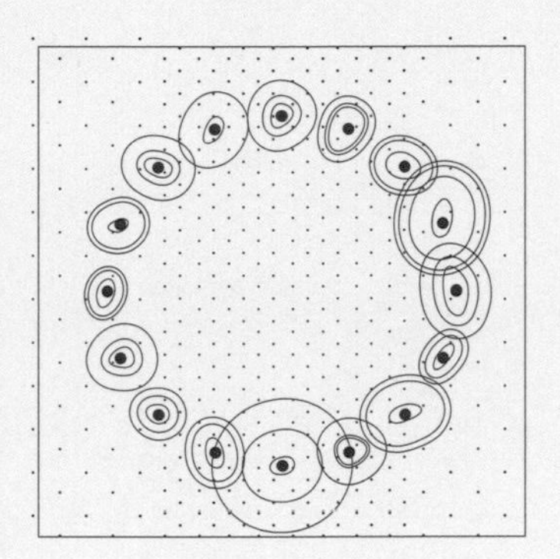

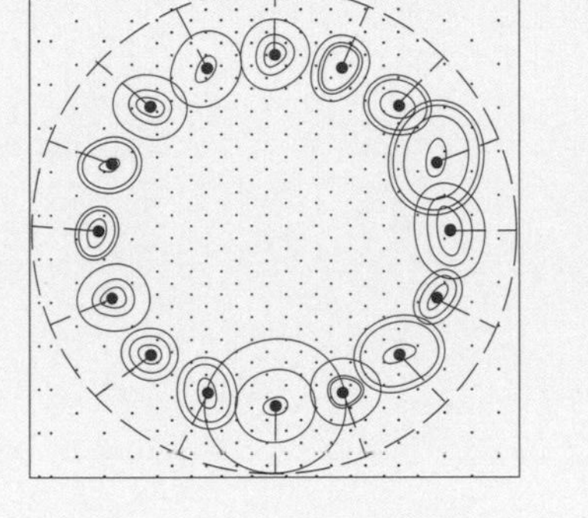

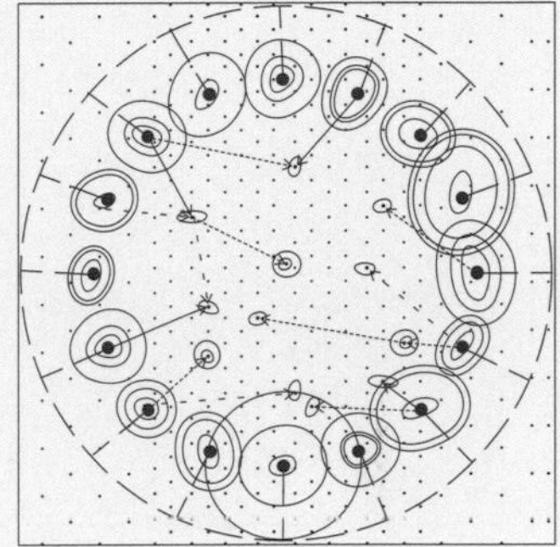

4

Navigate: arrange the points of life in a circle. This circle is a nonhierarchical framework for bringing together home ports. We'll call it the "portolan chart." It serves as a landmark, delimiting an area, throwing a net into the sea of living things.

5

Extrapolate the figures of the "habitual territories" and classify them by family or mode of trajectory, since following a route or disseminating a substance (pollen in the case of plants) is the fundamental mode by which living things inhabit the Earth. In future projects, this diagnostic step could make it possible to preestablish a land development program in a territory in order to take into account the trajectories of each species and human and nonhuman individual.

6

Trace the tracks of the living territories to the center of the map. The resulting structure or mesh is a living landscape. When tested on a territory, it provides a visual rendering of the way in which the points of life create space with their movements, their wandering, their intersection, their interlacing or their avoidance of each other, and their bifurcation. The landscape is constructed from the arrangement of the points and lines of the collected living things. The landscape is a woven assemblage.

MODEL III
LIVING LANDSCAPES

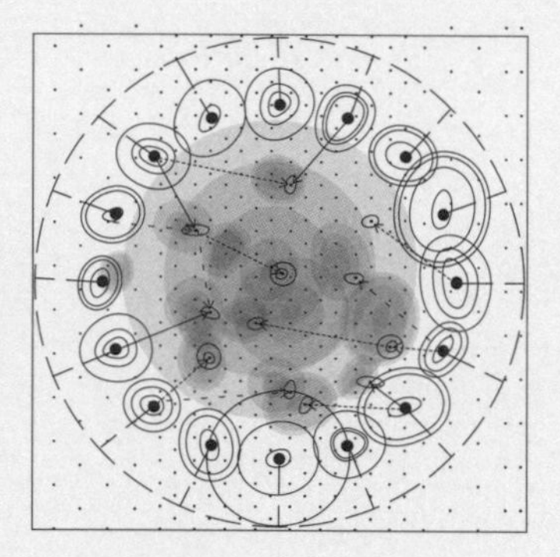

7
Extrapolate the forms of landscapes based on the paths of living things. This prospective step could be combined with a project of constructing space: a park, neighborhood, or public space. The routes, shapes, and consequences could be reworked to constitute topographies, paths, and spaces inspired by the habitual territories or dwelling places of those who live near the project's location.

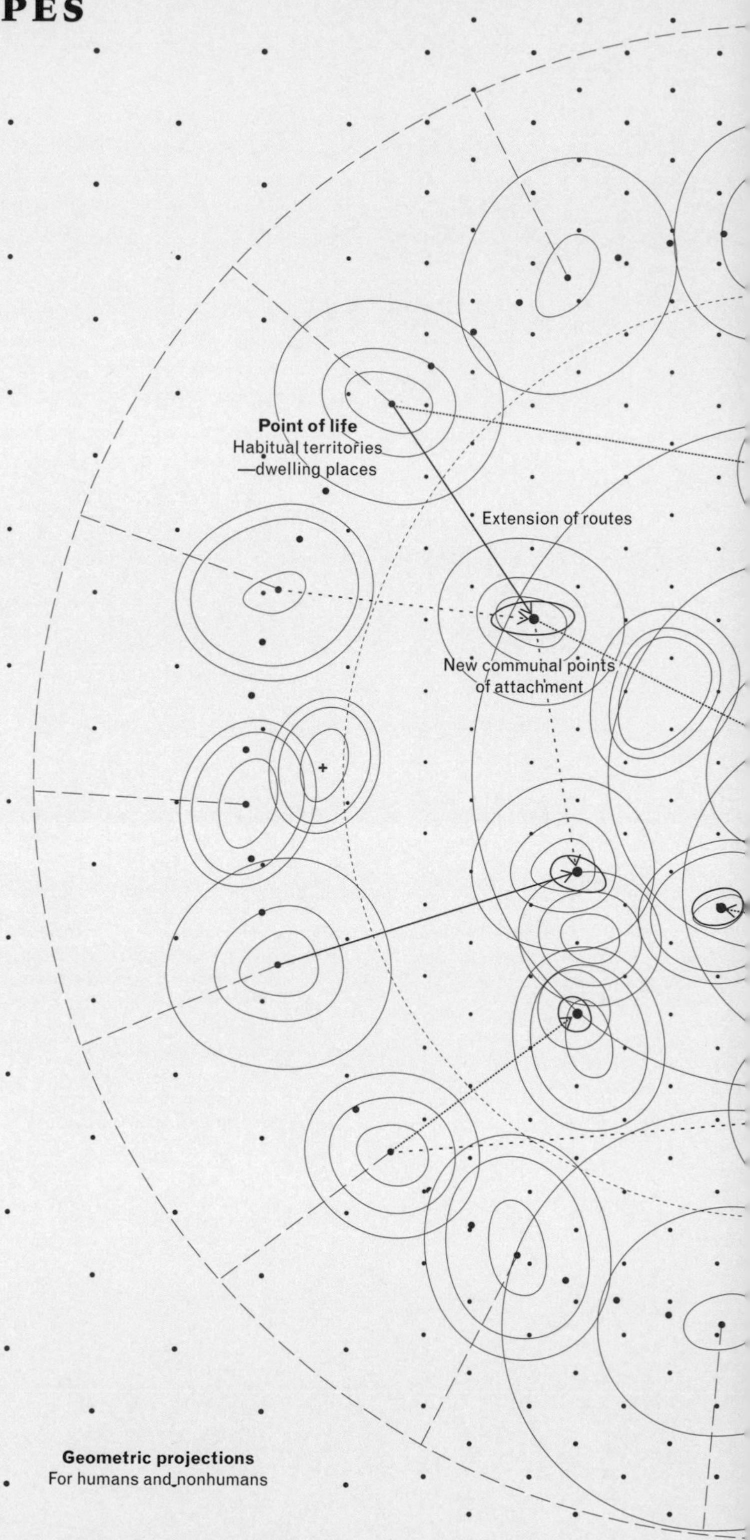

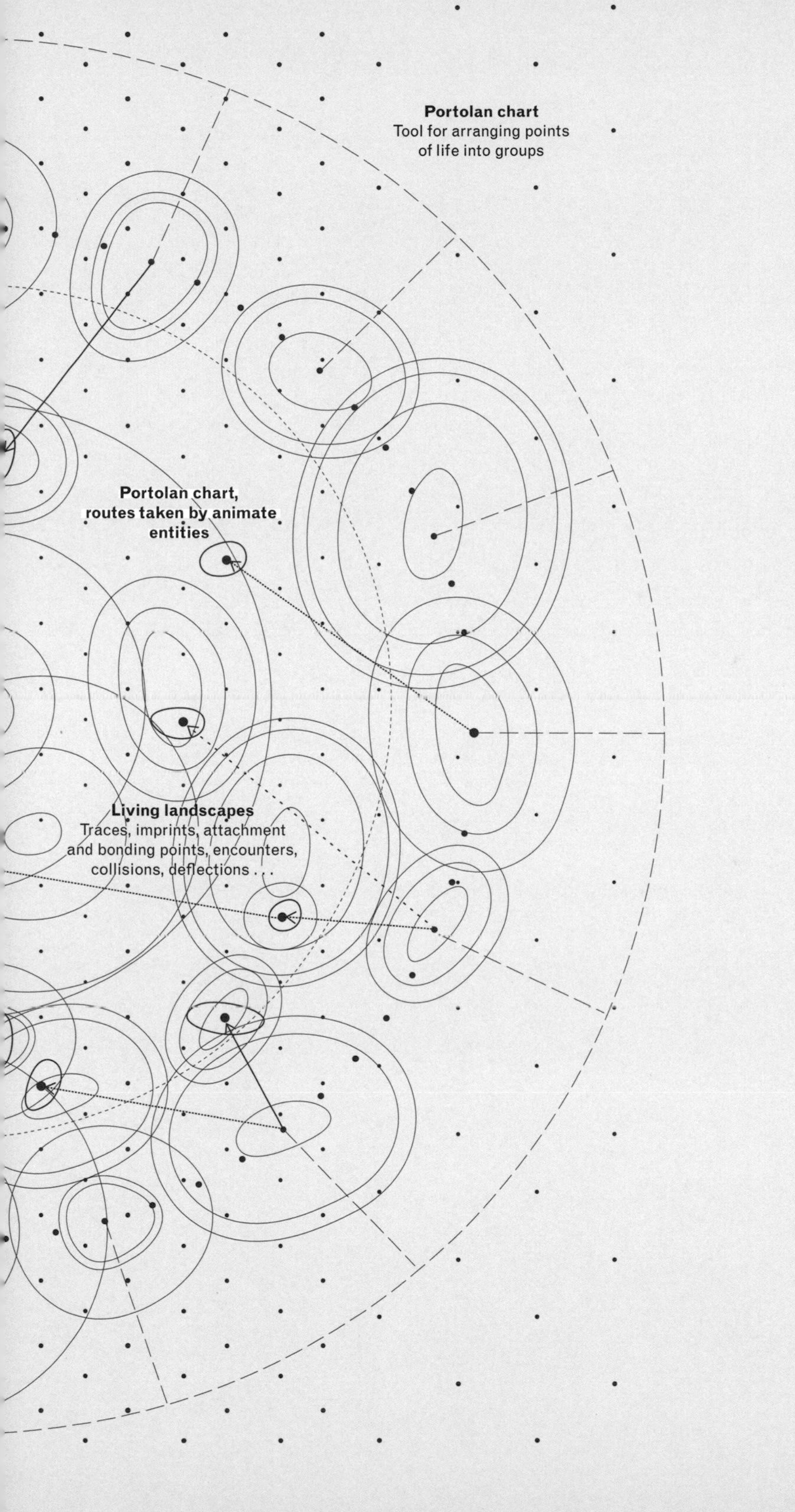
Portolan chart
Tool for arranging points
of life into groups
Portolan chart,
routes taken by animate
entities
Living landscapes
Traces, imprints, attachment
and bonding points, encounters,
collisions, deflections . . .

The task: to draw living things

The figure of the knot or flower: development of multiple nonconcentric loops or paths around a point

Suburban domestic cat

Suburban—teacher, 60 years old—Male

The figure of the network: extensive network of intangible connections with minimal physical imprint

Owl

Urban—self-employed (research), 33 years old —Female

The figure of backtracking: Returning to the same place, digging deeper into or, on the contrary, capturing a fragmented territory – possible fractalization of a territory

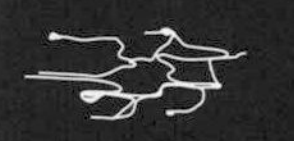

Boar

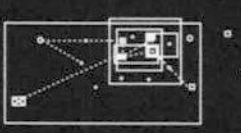

Urban—artist, 35 years old —Female

The figure of conquest: Rhizomatic development —expanding territory

Swift (bird)

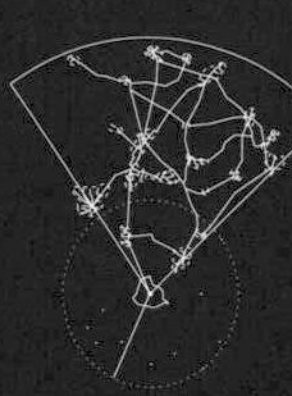

Suburban —salesperson, 25 years old—Male

MAP III LIVING LANDSCAPES

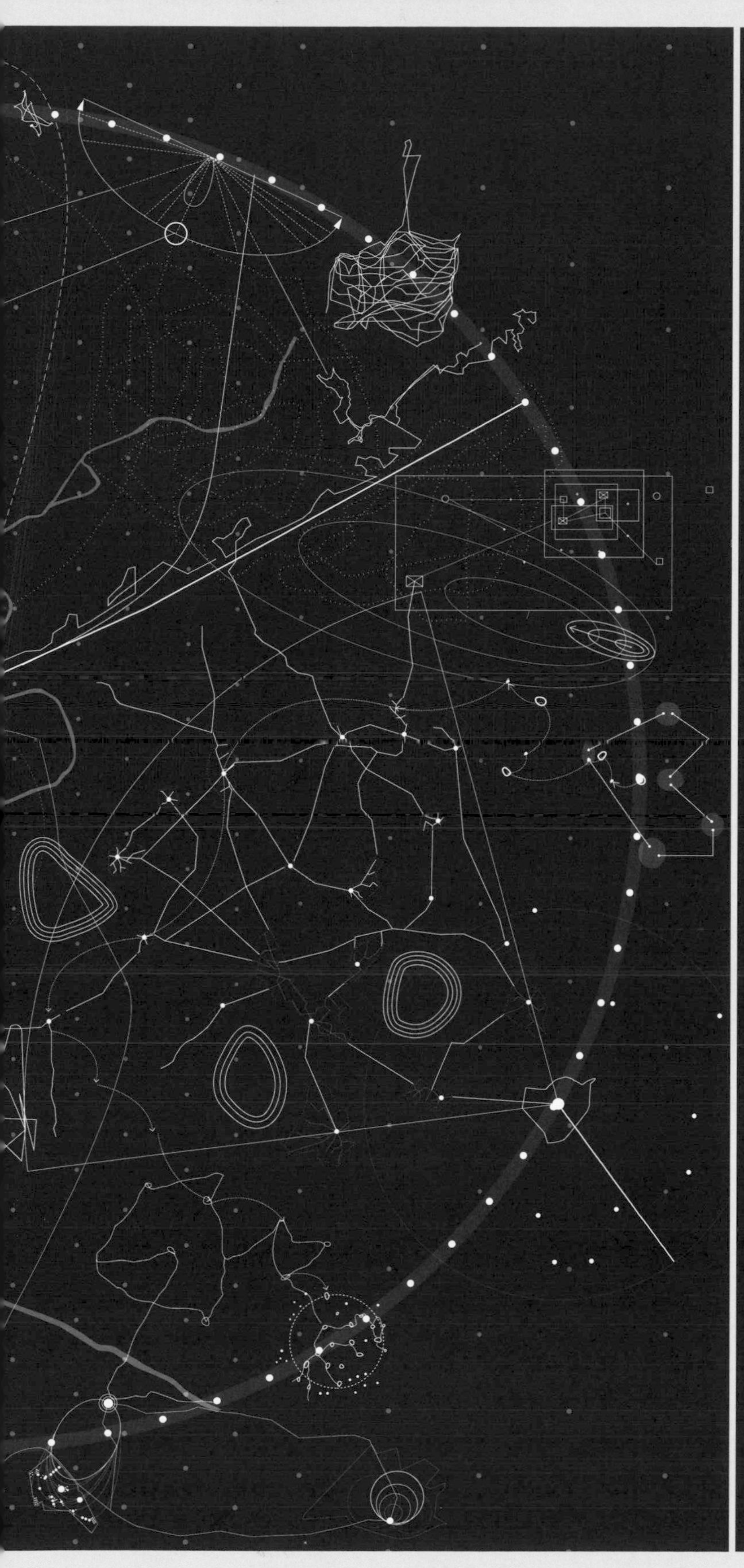

The figure of the egg:
Route and detours within a minimal pocket area

Ravens

Suburban—childcare worker at a maternal and pediatric health care center, 45 years old —Female

The figure of canvassing:
Exploitation of whole surfaces

Suburban—physical education teacher, 28 years old—Male

Mole

The figure of the furrow:
Development around one major axis

Suburban—manager, 30 years old—Male

Wild stone marten in suburban milieu

The figure of the leap:
Going from one micro-territory to another by borrowed means

Urban—permanent employee (architecture), 30 years old—Female

Microorganism

The figure of the pendulum:
Selective movement around a fixed axis

Ants

Urban—permanent and freelance employee (architecture), 30 years old—Male

TRANSFORMATION III
LIVING LANDSCAPES

From the grid . . .

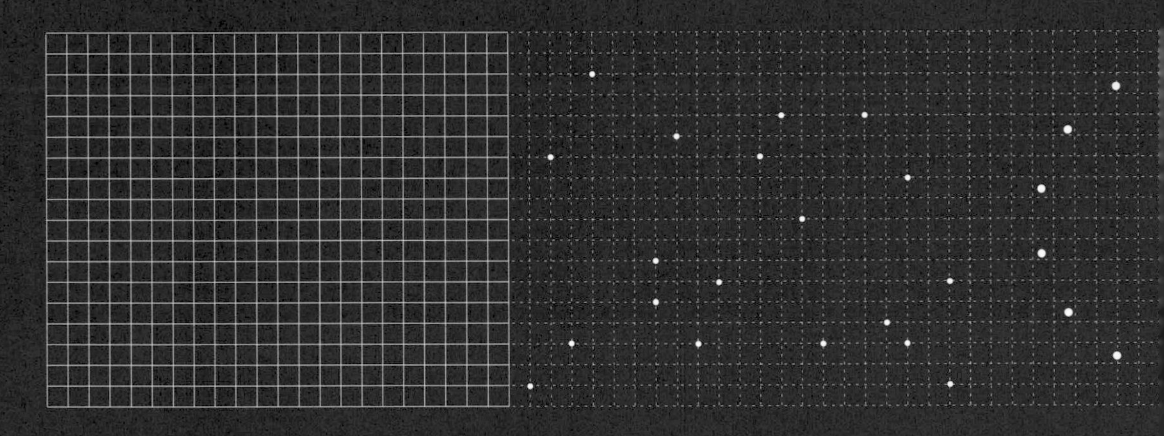

. . .to territories of living things.

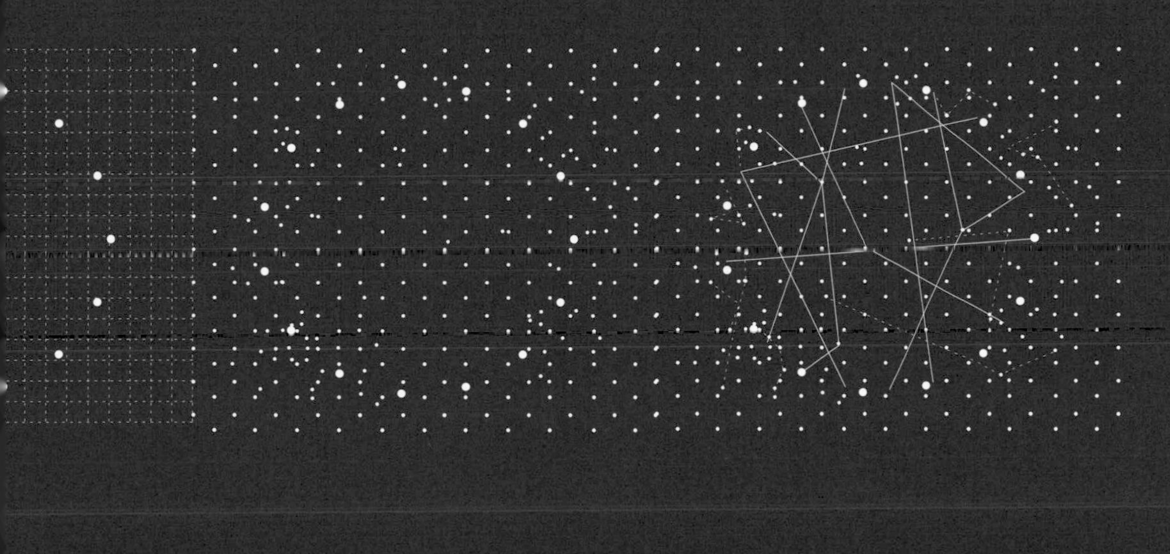

fig. 19 Map of the French-Spanish border, the Pyrenees

MODEL IV

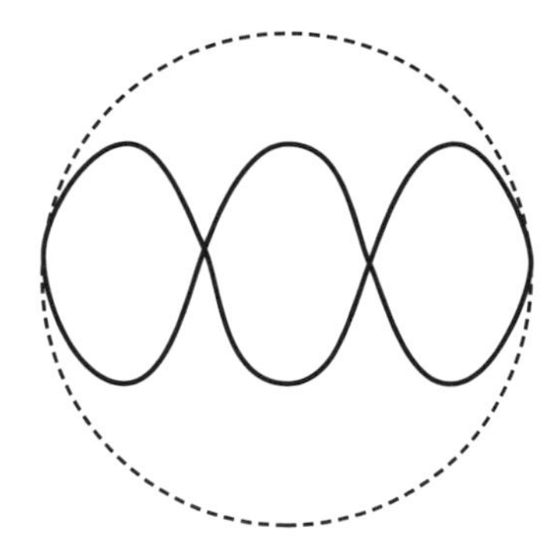

BORDERS

Demarcation

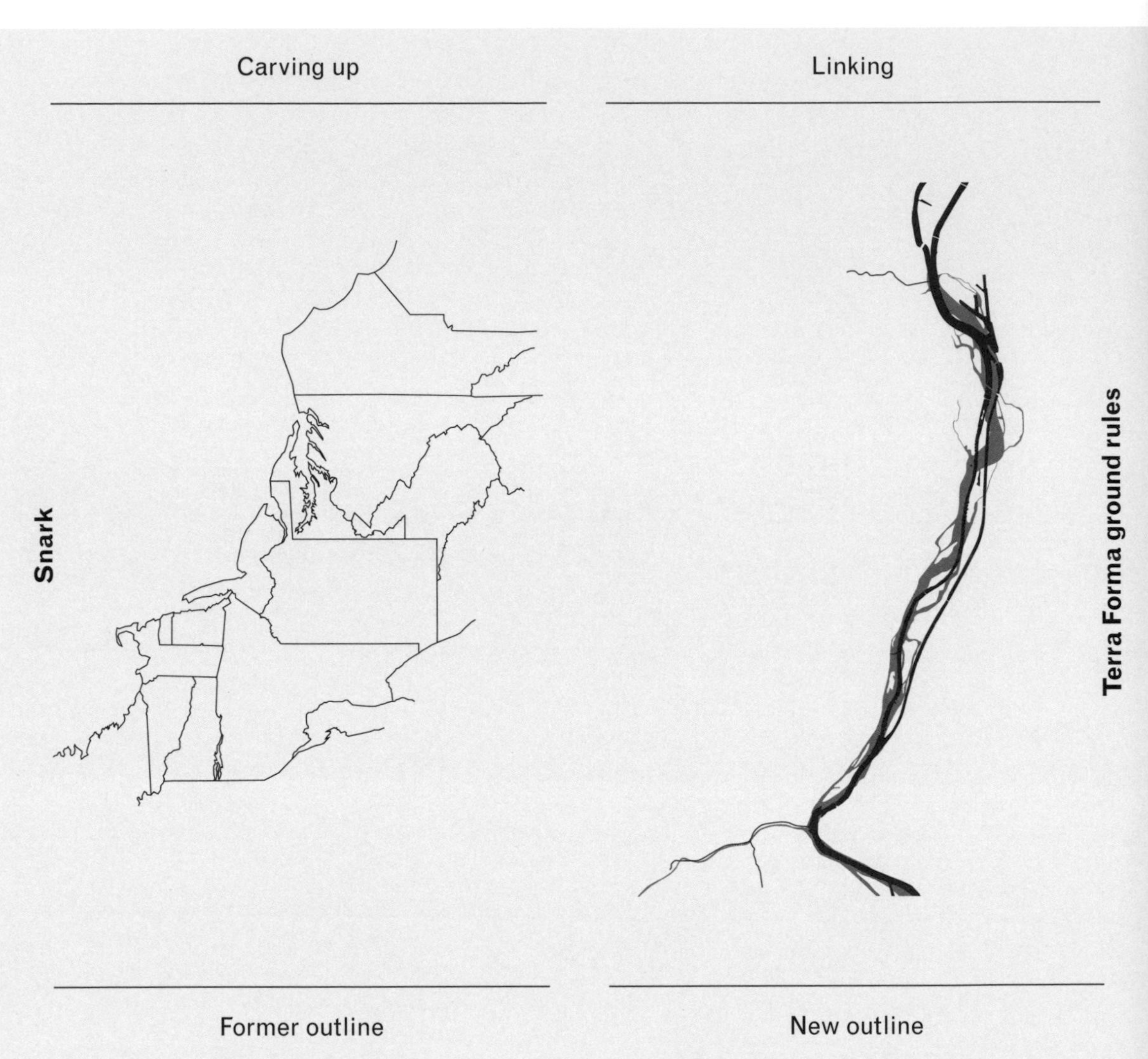

Map of the Rhône layering the outlines from the nineteenth and twentieth centuries (natural bed and artificial canal), south of Lyon.

BORDERS

Within a single landscape, living territories overlap, and any movement involves crossing thresholds, so that living things constantly evolve along, within, and across borders. You could say that the border is their milieu. In turn, every milieu is a border in a certain sense, that is, an interface and a zone of exchange. On the maps familiar to us, however, borders are lines, not spaces. Whether they mark the boundaries of villages or cantons, regions or nations, maps have played an essential role in the establishment and solidification of these initially arbitrary borders (even if they are sometimes supported by some physical reality), which are then gradually inscribed into reality, each line on a map having an impact on living things, their modes of existence, and their paths (fig. 19). The cartographic tradition's enthusiasm for borders, their physical establishment, and their visual representation by certain animal and human beings, can perhaps be explained by the need to create frames—a need that varies by society and species—in order to delimit a protective space, an envelope, even though this space constantly eludes us because we share it with others, and because it depends on others. These fictitious frames around us limit our movement, but they also hinder that of others—other species or other members of our own species. Since we take space with us wherever we go, it is inherent to the nature of these frames that they move and evolve with bodies in motion. Also, when our neighbors' fictitious frames collapse, they must cross the frames of others to survive. However, frames always end up collapsing because they are no more stable than space, a product of other living things, liable to shift according to the distant and nearby influences and effects of life. Our hypothesis is that borders are not fixed frames, but rather distortions of the transformation-as-creation of the planet by life, in favor of life, to make it habitable. The Borders model offers a metaphorical-metamorphic visual representation of this twisting of space by living things. We attempt to curve the immobile line of the border by literally bending it on the map.

We know that world maps give the impression of a flattened world with the edges distant from each other (the map generally being cut in the middle of the Pacific), and whose land is divided by as many borders as there are nation-states. But the Earth is a single, continuous, curving surface. This continuous, inhabited border, a Möbius strip, without beginning or end, without interior or exterior, offers us an interesting graphic structure. The Möbius model represents a seamless, continuous but heterogeneous territory whose shape, size, and nature are constantly changing (see model pages). A sort of labyrinth-world and anti-atlas, this tool allows for the perpetual experience of a limit within a single territory. In fact, for us this whole book is an anti-atlas because ultimately what we have are not maps, living things, and a world separate from them, but rather efforts at putting the same reality into perspective—this land/map/group of living things that we approach from a different point of view in each model. One chapter does this from the perspective of the monadological inclusion of the world in the living thing, another from the perspective of the living thing, and this one from the perspective of friction and tectonics between different living things.

Through their presence, their growth, and their movements, living things can generate varied effects—erosion, concretion, construction. These effects serve a purpose: they allow living things to manage the relationships between their body and others' bodies by transforming mineral or plant matter (in the case of humans), by making organic or olfactory deposits (for certain animals), or by emitting chemical signals (for plants).

The question then arises: what are the methods and conditions for this co-construction of common space, and for cohabitation with other living things? If borders are where the effects of living things overlap, and where the reach of their movements is negotiated, the proposed model could help us draw the map of a geopolitics of living things that several philosophers have called for.[35] To inhabit a border is to inhabit interfaces where sometimes divergent interests crystallize, where distinct territories are linked by their shared edges. These edges are the site of exchange and dynamics, conflict and sustained negotiation. As a first step, we will map the present forces at play: the border locations of the territory we hope to analyze.

The resulting map is based on the Pyrenees, a transborder territory between France and Spain for which the available cartographic data is ample (see map pages). This territory is also of interest to us because it contains different types of physical and cultural borders, which the map forces us to cross: Andorra, a sovereign landlocked state of 468 square kilometers and a tax haven; a mountain range and the sea, which are natural borders; areas of conflict between France and Spain concerning water usage and around the presence of a language and culture of its own in Catalonia; the discontinuity of the infrastructure (the dimensions of train rails differ between Spain and France); cohabitation between humans and large predators; and the twin villages on both sides of the border, which were border posts. By peeling back top layer, we create a surface map that forms a continuity, a skin map of "mountains-water," to cite the Chinese concept of landscape—a lived landscape.[36]

---

35 See in particular Stengers, *In Catastrophic Times*; Anna Tsing, *The Mushroom at the End of the World*; Coccia, *The Life of Plants*; Morizot, *Les Diplomates*.
36 This concept is elaborated upon in chapter 2 of François Jullien, *Living Off Landscape or the Unthought-of in Reason*, trans. Pedro Rodriguez (London: Rowman & Littlefield, 2018).

In this model, borders are classified into three types that have three distinct morphologies, natures, and scales (see model pages). Border lines can be ridges or barriers; we will call them *limes*, a Latin term for fortifications (fig. 20). In our model, they are the narrowest loop of the ribbon. But however thin it may be, this line still has thickness and materiality. Environments, or ecotones, are interface spaces endowed with a definition and an identity of their own despite the fact that their function is to position things in relation to each other. This is true of the edges of forests, for example, which have their own ecology (fig. 21). The outer reaches have another scale, a territorial one. They refer to an outside world rather than an interface (fig. 22). As a second step, we identified four roles played by borders: they are a place for exchanging goods and ideas (exchange zone), regulating and controlling (regulation zone), transforming the identity and value of beings and things (metamorphic zone), and fighting or taking refuge (conflict zone). Added to these types is another kind of place: the inclusive border, which operates under a specific system and is governed by specific laws, such as airports, embassies, and de facto borders (e.g., Lampedusa, which has become a territorial border since the erasure of national limits in the Schengen area).

Border negotiations here are no longer so much between humans as between a mountain and its entities on the one hand and human habitats on the other, tacitly linked on either side of national borders. The sum of its variations and the diversity of its landscapes will constitute the particularity of this territory. In this sense, this model is comparable to a DNA sequence. Its design is also based on a double helix. On the one hand, the helix of human constructions of borders , where we find crossing infrastructures (which go beyond the natural border, crossing ravines or seas), border posts, international zones (airports), gates and thresholds of the city all the way down to domestic

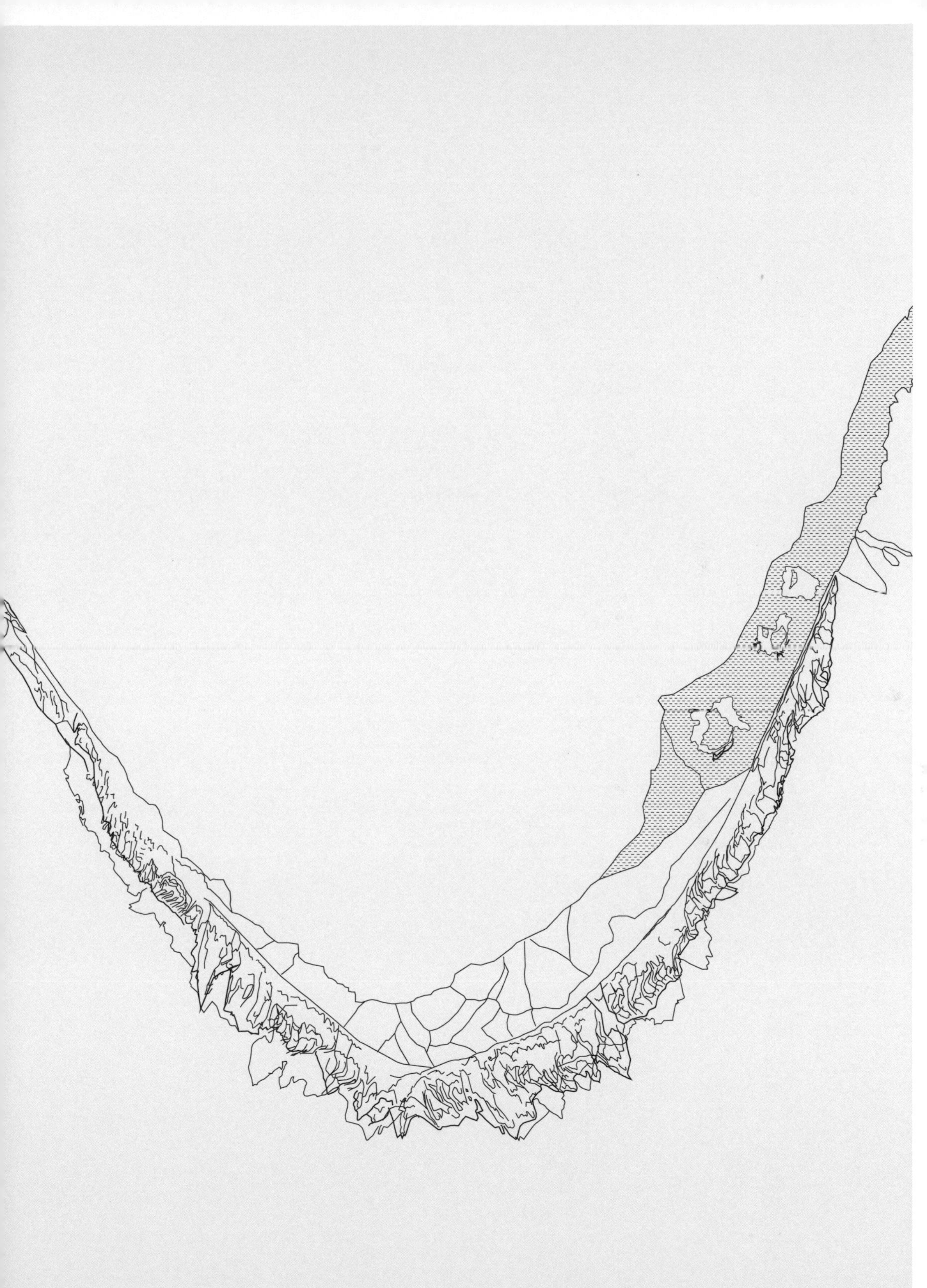

Fig. 20 *Limes*

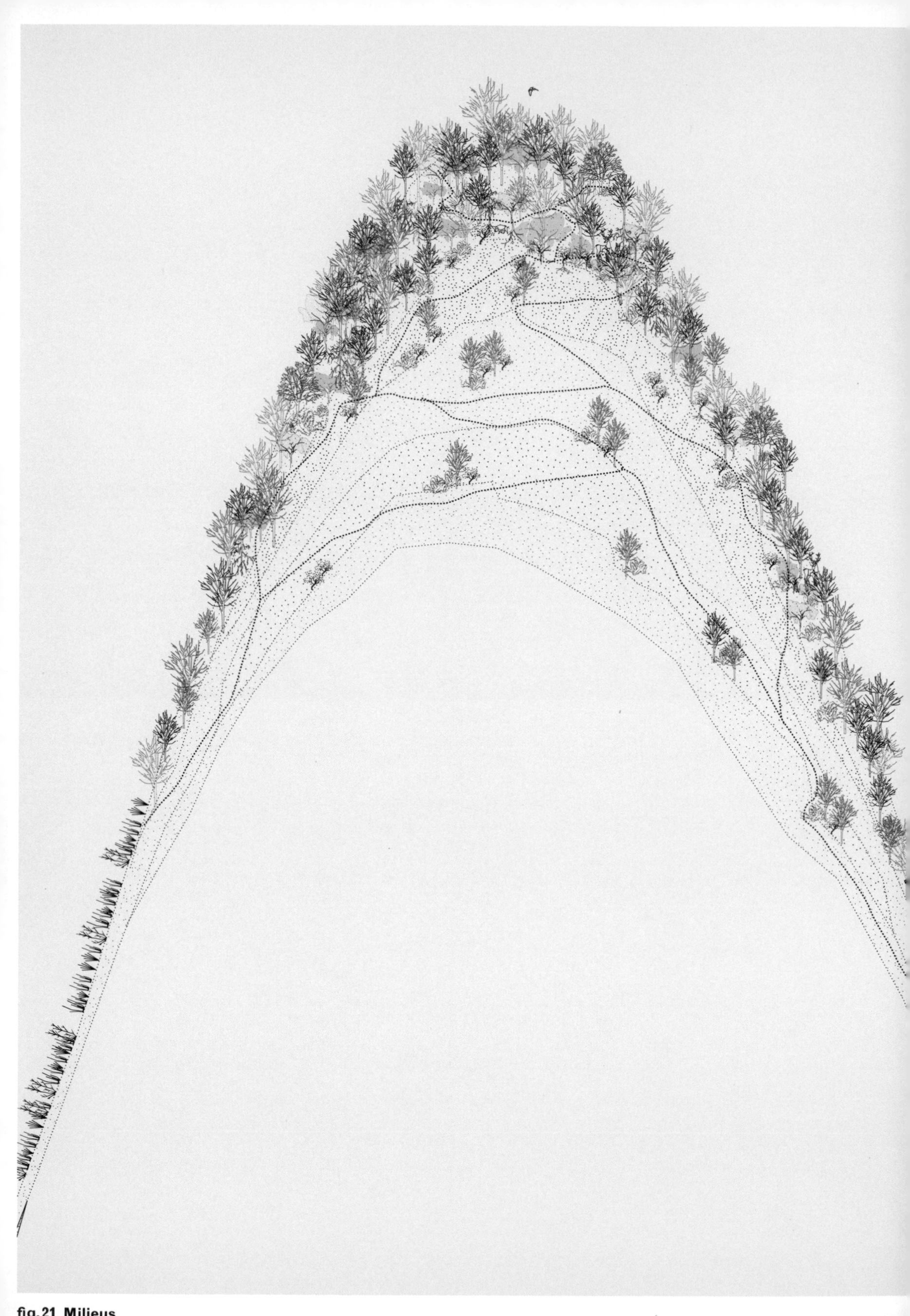

fig. 21 Milieus

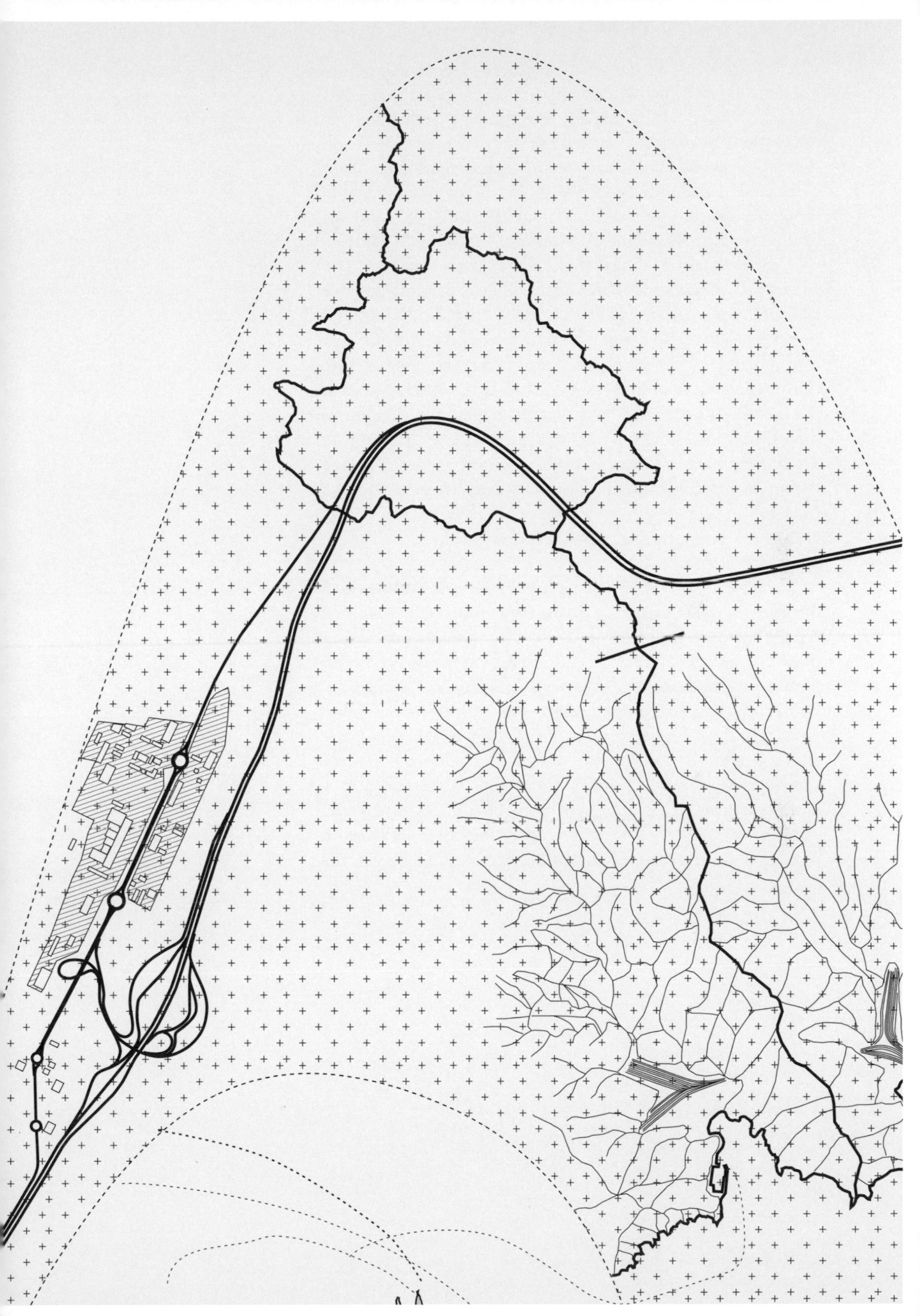

ig. 22 Outer reaches

spaces (thresholds of houses, boundaries or gradations from public to private), no-man's lands (areas of conflict over maintaining a border), and border regions (regions that benefit from the border economically or geographically). On the other hand, the helix of nonhuman border constructions : mountains, peaks, cliffs, and ridges, which by the action of erosion turn into beaches when they touch the sea, then into wetlands, ponds, or peat bogs when they come into contact with arable land; from this low vegetation, forests gradually develop that maintain edge zones, ecotones with abundant biodiversity between two environments, with fields and meadows inhabited by animal populations that also maintain territories by mapping out their own moving olfactory ecotone. Each helix is intertwined with the other and both unfold from the *limes* to the outer reaches. If the human line seems to have sharper breaks than the nonhuman line, the distinction will not always be so pronounced, depending on the territory studied with the model. The two curves form a single ribbon, which can also be read face to face from top to bottom—an area where properties are sometimes exchanged. Presented in this way, a border is no longer a line but a shore between worlds.

Now that we know that every border is not a line but a thickness, what does this mean for the cohabitation of actors? What phenomena bring about these sites of conflict or adaptation? The model explores this question with the use of oscillographic diagrams (fig. 23). The shape of the helix expresses the border's ambivalence as both a separating space and a common space. The upper and lower part of each twist of the helix is the starting point for the interests of each border place in its relationship with the area facing it, with whom it shares concerns and edges. The idea is to be able to map the forces at play and potentially problematic situations at a glance, in order to try to sketch out avenues for reconciliation. The oscillogram records

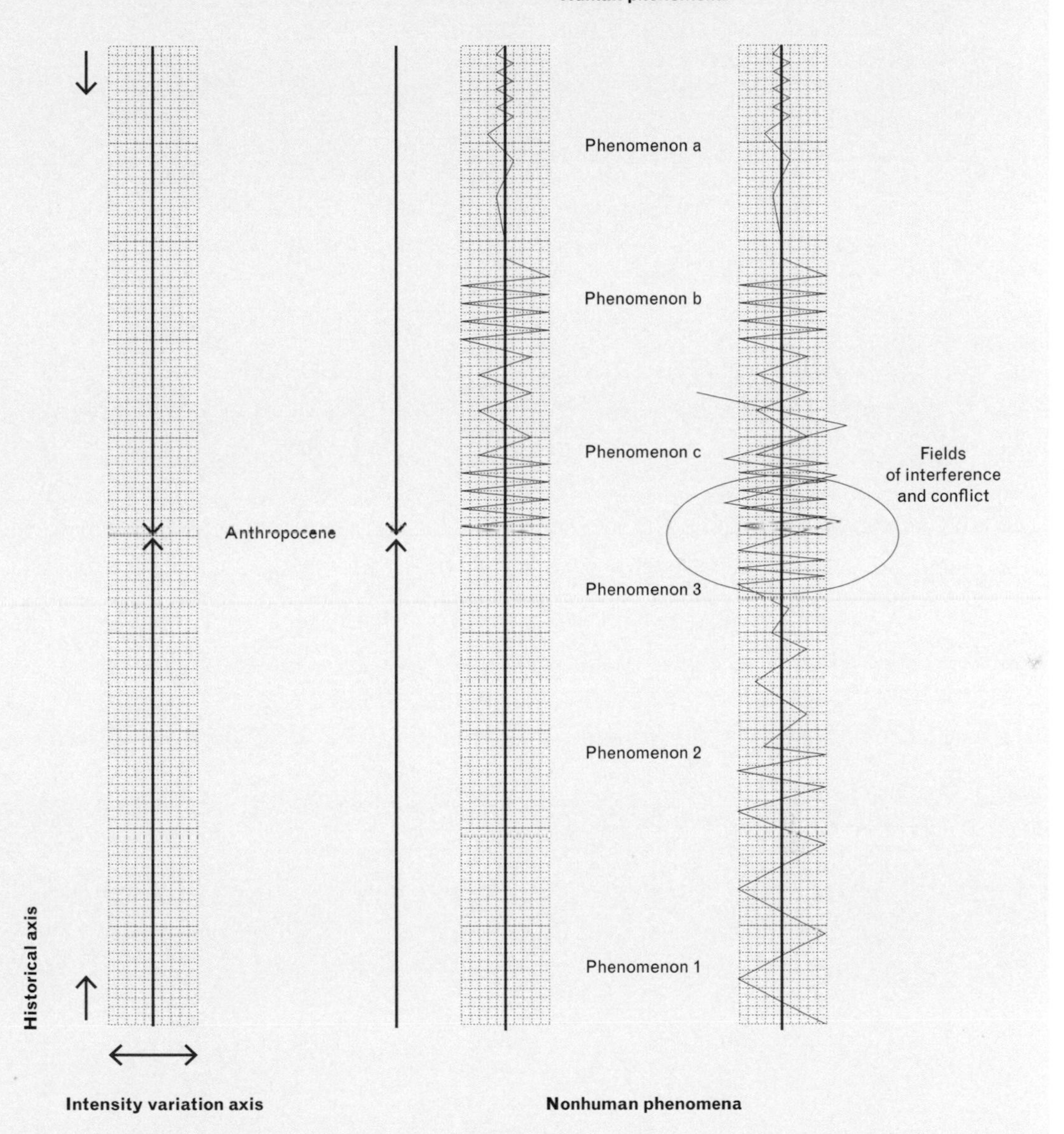

g. 23 Oscillographic diagrams Management of cohabitation conflicts on the border

the pulses of living things, and human–nonhuman polarity; it attempts to measure the combined scale of needs and resource consumption, which is often unbalanced. It could be developed as a timeline signaling the peaks and valleys in the intensity of the phenomena that caused the territories to mutate—phenomena of both natural and human origins, mixing and multiplying to explain the climaxes we see.

MODEL IV BORDERS

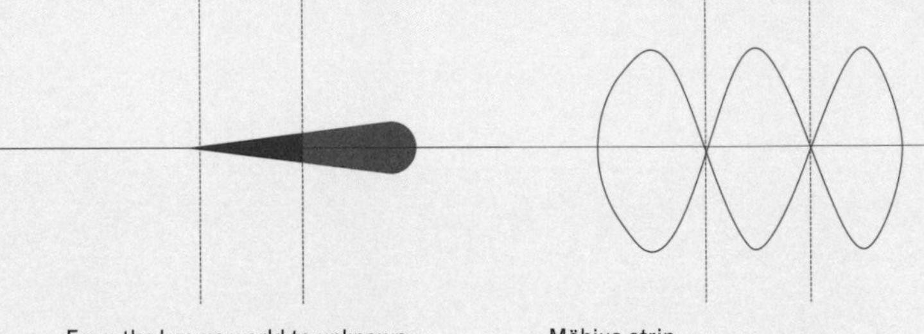

From the known world to unknown frontiers

Möbius strip

1

Identify and provide a morphological classification for a given territory's types of borders: *limes* or milieux or outer reaches. *Limes* correspond to known borders, whereas outer reaches designate unknown worlds. Milieus lie between these two extremes, corresponding to interface territories.

2

Manipulate the border line to obtain a continuous and infinite line. We thus obtain a double helix that, inspired by the Möbius strip, curves over on itself on either end of the chain. The mapped world is continuous, without beginning or end; we cannot reach its edges without finding ourselves in the midst of the loop again, somewhere on the continuum of a sequence of places.

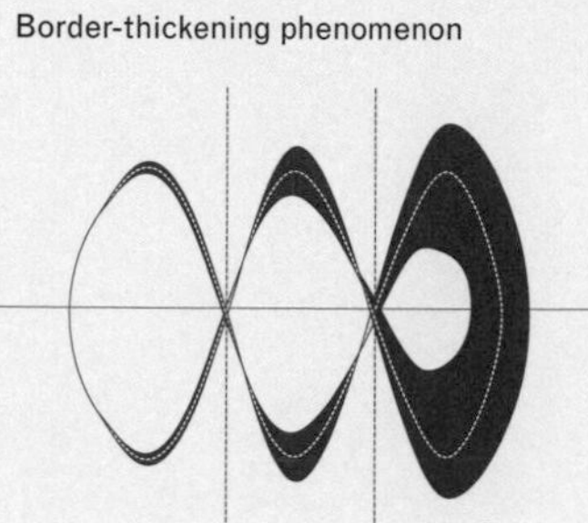

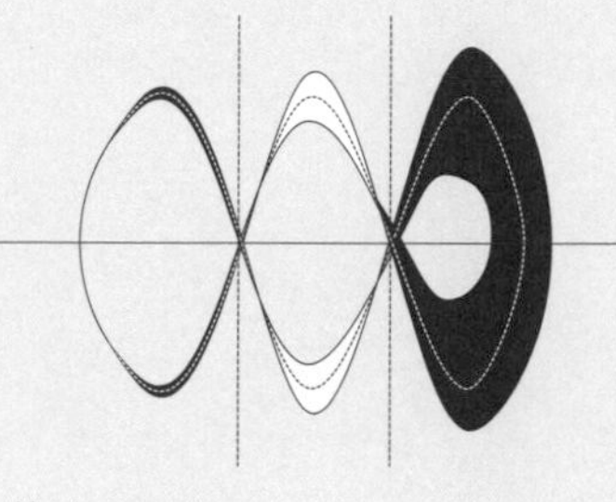

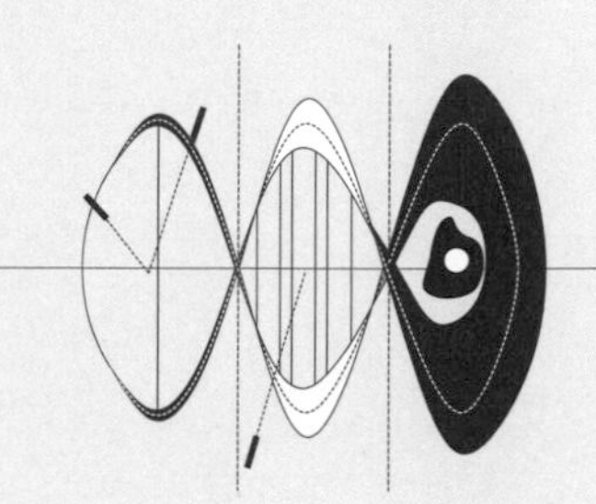

3

Adjust the thickness of the strip according to the morphological classification determined in the first step, with *limes* consisting mainly of narrow or tenuous places, milieus of wide or thick spaces, and the outer reaches of vast territories with an inaccessible horizon.

4

Distribute the border spaces on one strand of the helix depending on their human or nonhuman affiliation. The strands cross and merge into each other; the human and nonhuman worlds sometimes face each other and sometimes intermingle. Map the materiality of these border worlds according to the principles of tapestry (an axonometric representation, in which the features of certain entities are exaggerated to emphasize their importance).

5

Based on the territory studied, identify the types of relationships between the strands of the helix: exchange zone, regulation zone, metamorphic zone, or conflict zone. Place them on the map on the lines that stretch between the strands.

MODEL IV BORDERS

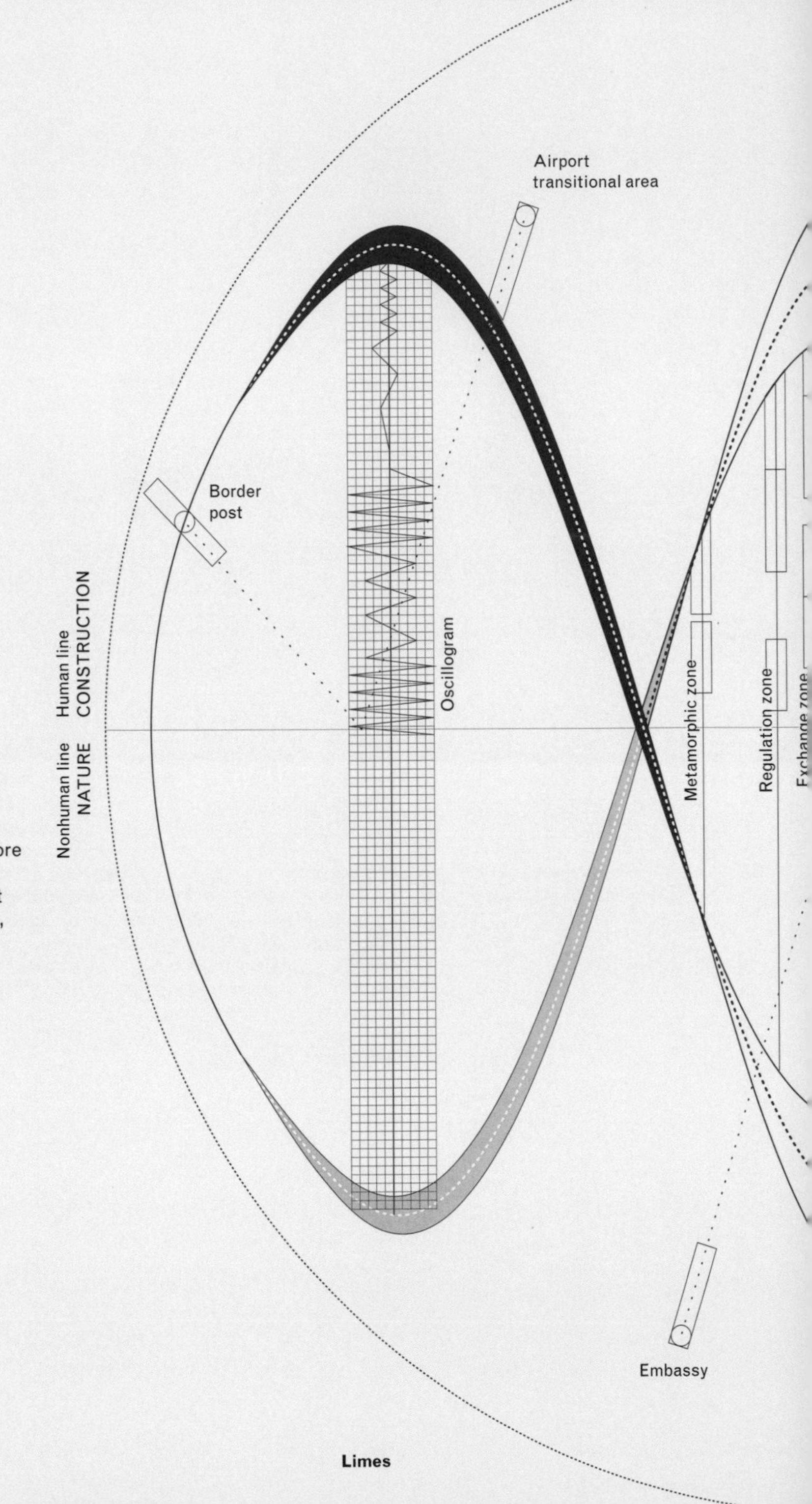

6

For each of the relationships identified, establish the oscillogram of the balance of power. Variations in intensity characterize the relations in a more or less marked way, locating and revealing the invisible forces that fabricate the borders: exchanges, regulations, metamorphoses, or conflicts.

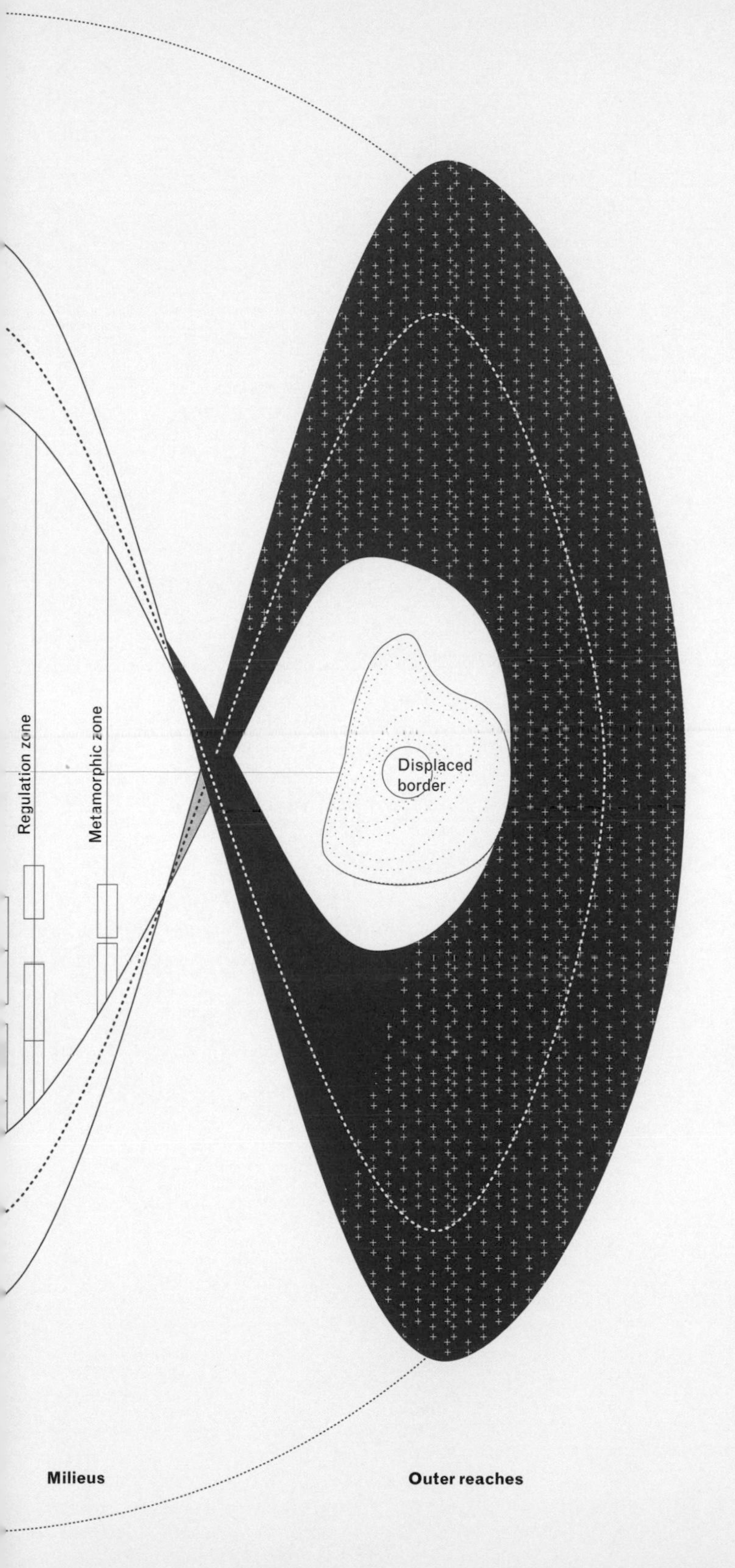

7

Place “ancillary” or neutral territories outside of the helix, along with territories that could be sites of reconciliation: embassies, border-territories (e.g., islands).

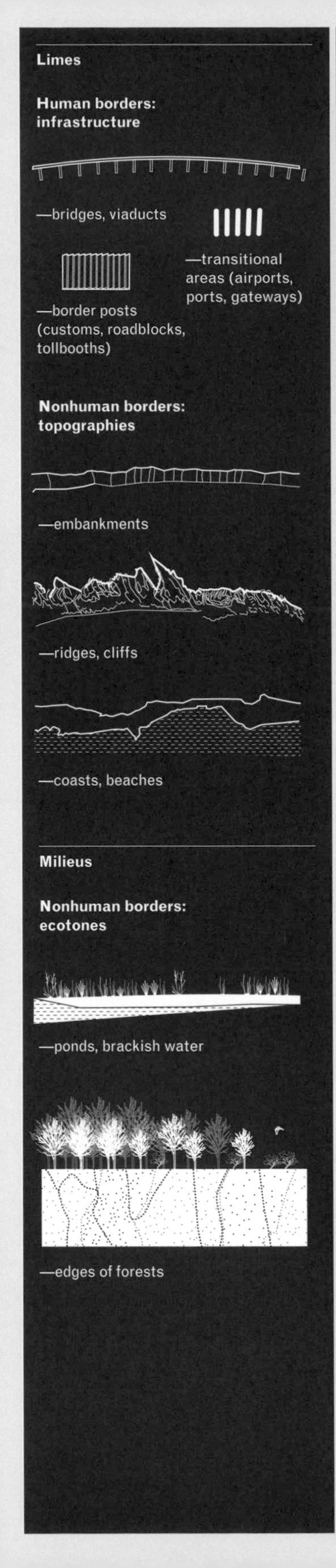

MAP IV BORDERS

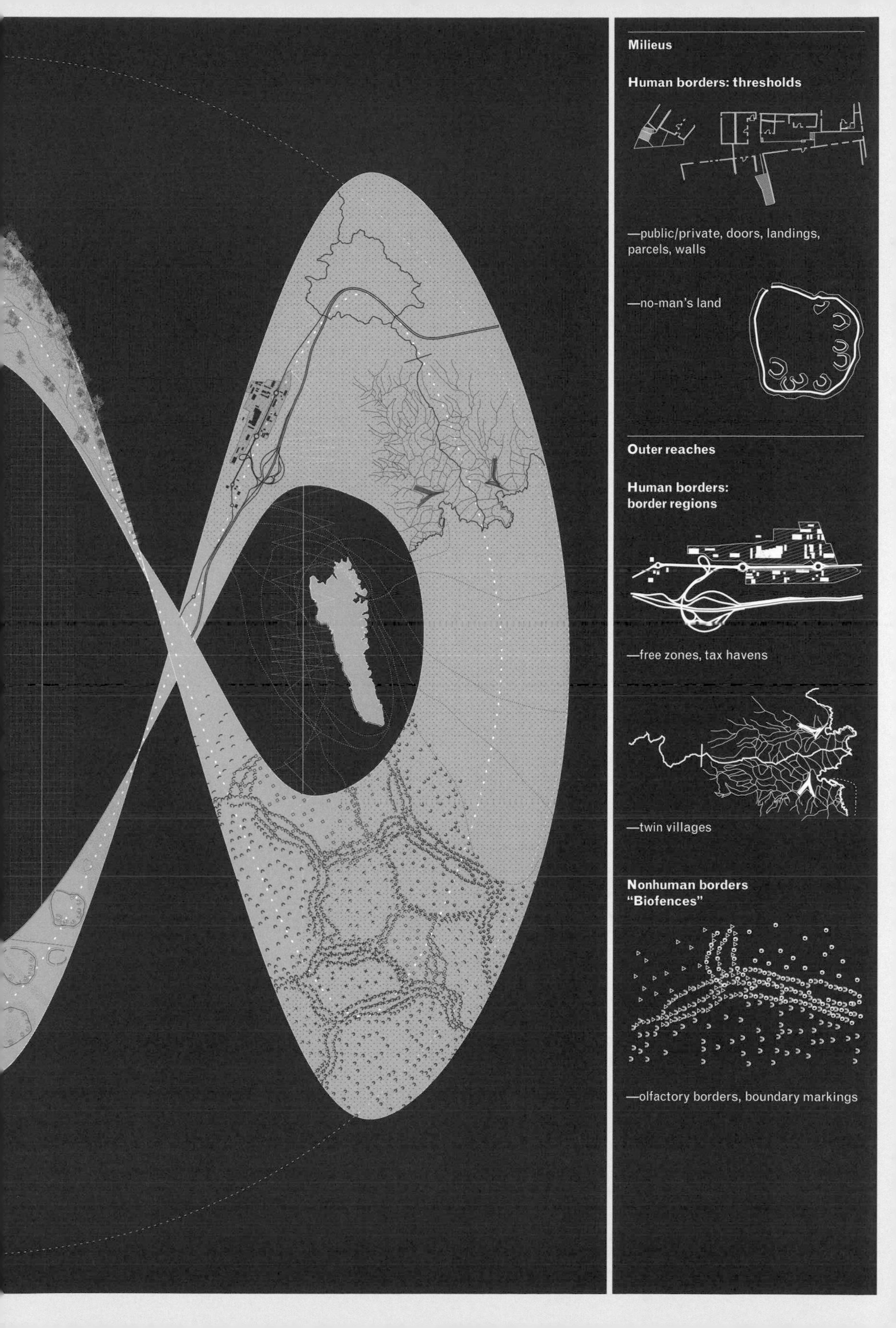
Milieus
Human borders: thresholds
—public/private, doors, landings, parcels, walls
—no-man's land
Outer reaches
Human borders: border regions
—free zones, tax havens
—twin villages
Nonhuman borders "Biofences"
—olfactory borders, boundary markings

TRANSFORMATION IV BORDERS

Borders . . .

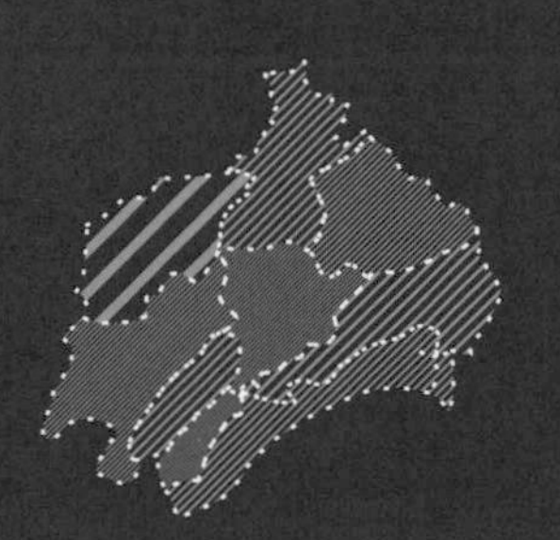

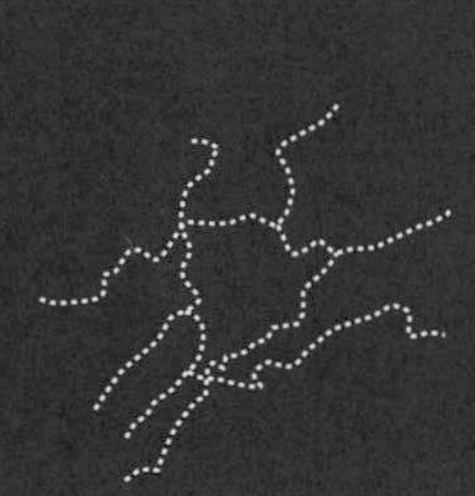

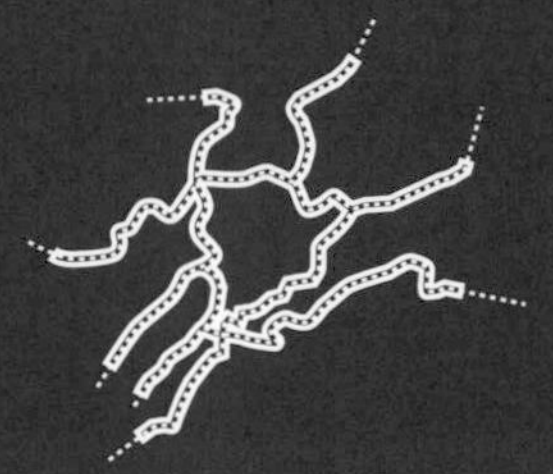

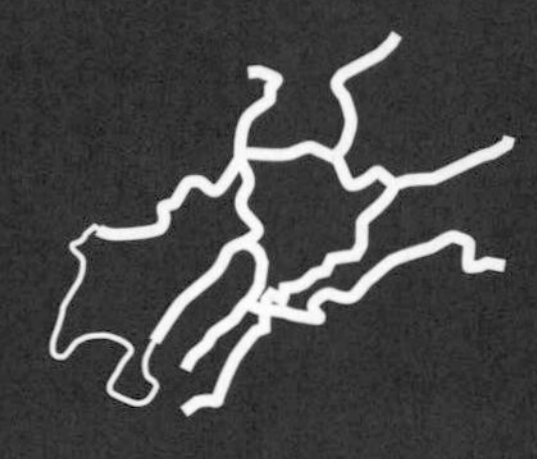

. . .in a continuous world.

fig. 24 Building map, Paris-Nord

MODEL V

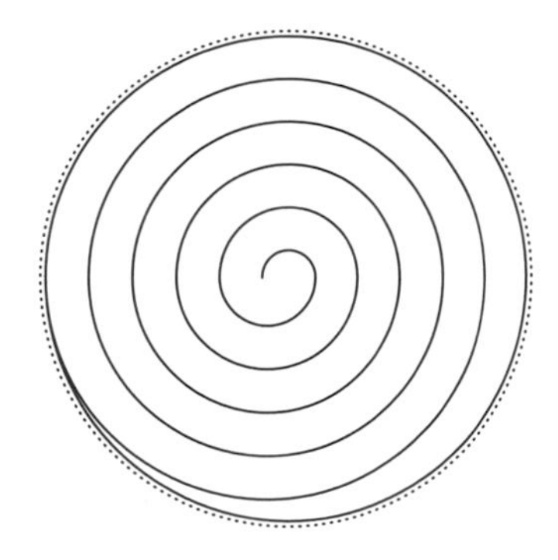

SPACE-TIME

Measuring

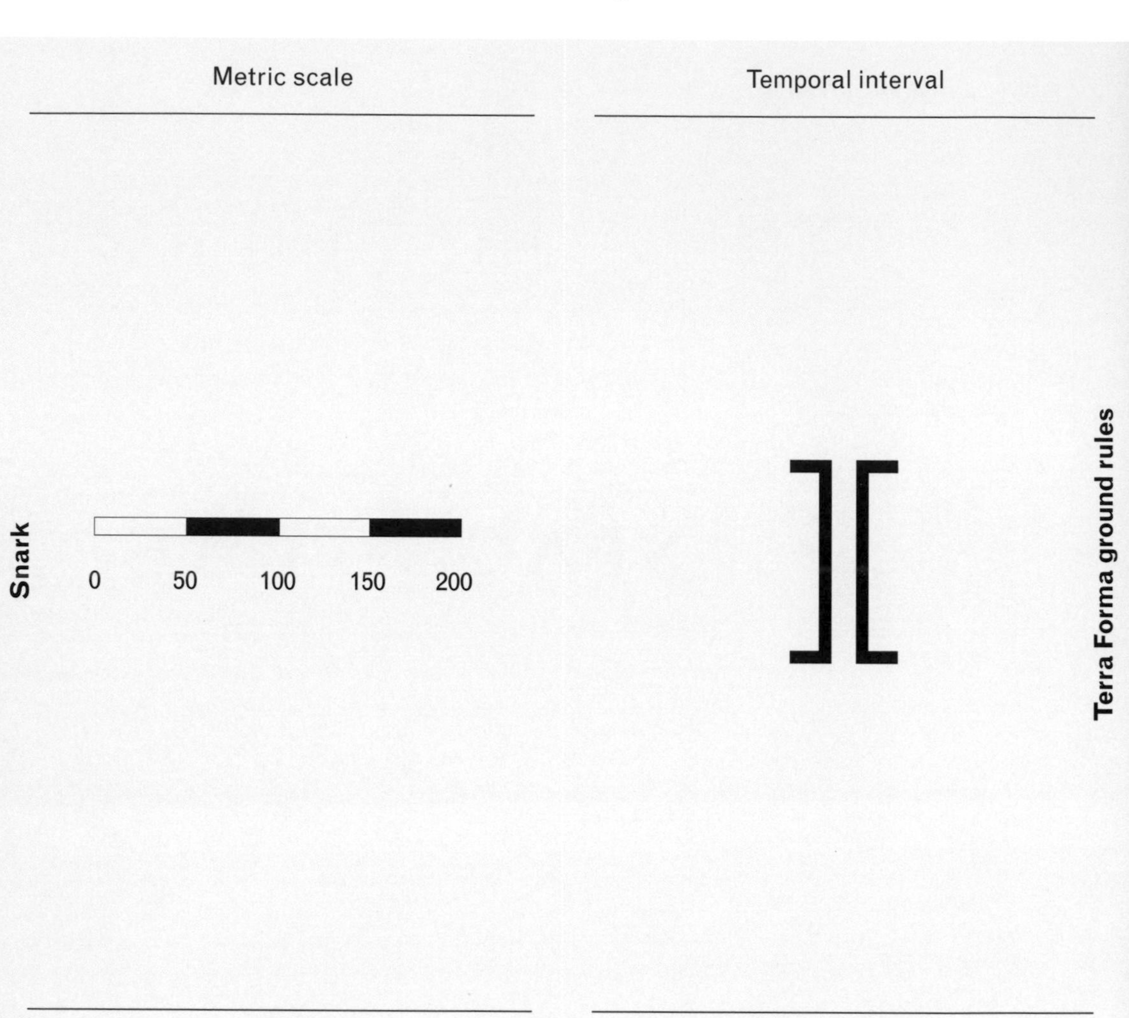

Metric scale
Temporal interval
Snark
0
50
100
150
200
Terra Forma ground rules
Former scale
New scale

SPACE-TIME

In the past, it was possible to program silent periods on jukeboxes. Shared times of silence, collective blanks. We still find blank spaces like this on the margins of the city, at its interstices, outside the metropolitan narrative.[37] But they tend to disappear. The formulation "24/7" is a demand for continuous time, permanent and smooth operations, detached from the impediments of material and space.[38] Hence the tendency to "overschedule," to saturate our spaces and calendars. As a new standard, however, this continuous time is not sustainable.[39] The desire for periods of silence and rest drives the exploration of those rare spaces that so far have escaped the planning that now determines the rhythm of various urban spaces (fig. 24). These parallel temporalities invite us to reflect on the connection of space and time in the hyperinscribed and hyperdetermined spaces of the city. In the following, our drawings will testify to this overscheduling. The text will express the desire for silence. Together, they seek to restore a sort of plasticity to space-time by helping us to situate ourselves between the poles of acceleration and deceleration.

Like landscapes and borders, space-times are shared, polyphonic spaces. To grasp them, we had to give up on reconciling conflicting times and reducing their different units to a

---

37 The writer Philippe Vasset has made these blank spaces one of the subjects of his work. See especially *Un livre blanc: Récit avec cartes* (Paris: Fayard, 2007) and *La Conjuration* (Paris: Fayard, 2013).

38 The entry of nature into the city helps to fight this demand. The ambition to create a "fertile city" as a political response makes it possible to use the time of living things, which is irreducible, and the forms of nature, which are uncontrollable, as a means of resistance. Parks, as places for strolling and wandering, are emblems of this desire to slow down. They are places where time expands.

39 See, e.g., Jonathan Crary's denunciation of this concept in *24/7: Late Capitalism and the Ends of Sleep* (London: Verso, 2014).

single tempo. The Space-Time model offers an alternative to the forced interlocking of conflicting times, which can lead to blockage, dislocation, and even spatial and social rupture. When we tried to imagine a tool that could give plasticity back to time, we were inspired by the modifications to space brought about by seasonal, climate, and geological phenomena: the rhythm of the tides, floods that temporarily hide pieces of land, the seasonality of forests, and their cycles of metamorphosis. Thinking of time as a geographical element—a river, for example—allows us to think about its bed, its minimal footprint, but also its capacity for spatial and temporal extension. A space-time's impact on the world is defined by its footprint and duration.

What is the spatial echo of time? How should a particular space-time be represented? What attributes would it have? Starting from the hypothesis that time cannot be abstracted from the space in which it is lived, we consider it as a substance, a material endowed with a form and physical properties. The model's aim is then to visually represent the correspondences, coincidences, and frictions between space and time in order to move from planning to a city of "potential times." We thus move from a single clock, or a multitude of synchronized clocks, to a desynchronization of times. In place of a universal dial, the model combines and layers of different tempos—known in contemporary music as polyrhythms.

The graphic matrix of the model is a score that opens up many interpretations and invites us to anticipate, bypass, deviate, accelerate, delay, model, and resume. Drawing inspiration from musical and choreographic notation, it seeks to visually represent a shared, manifold time (fig. 25). We developed a shorthand—or notation—from an elementary unit, the point, based on several parameters to allow us to perform a sort of cartographic stenography. This notation combines the notions of duration and footprint, which are equivalent to the space-time's

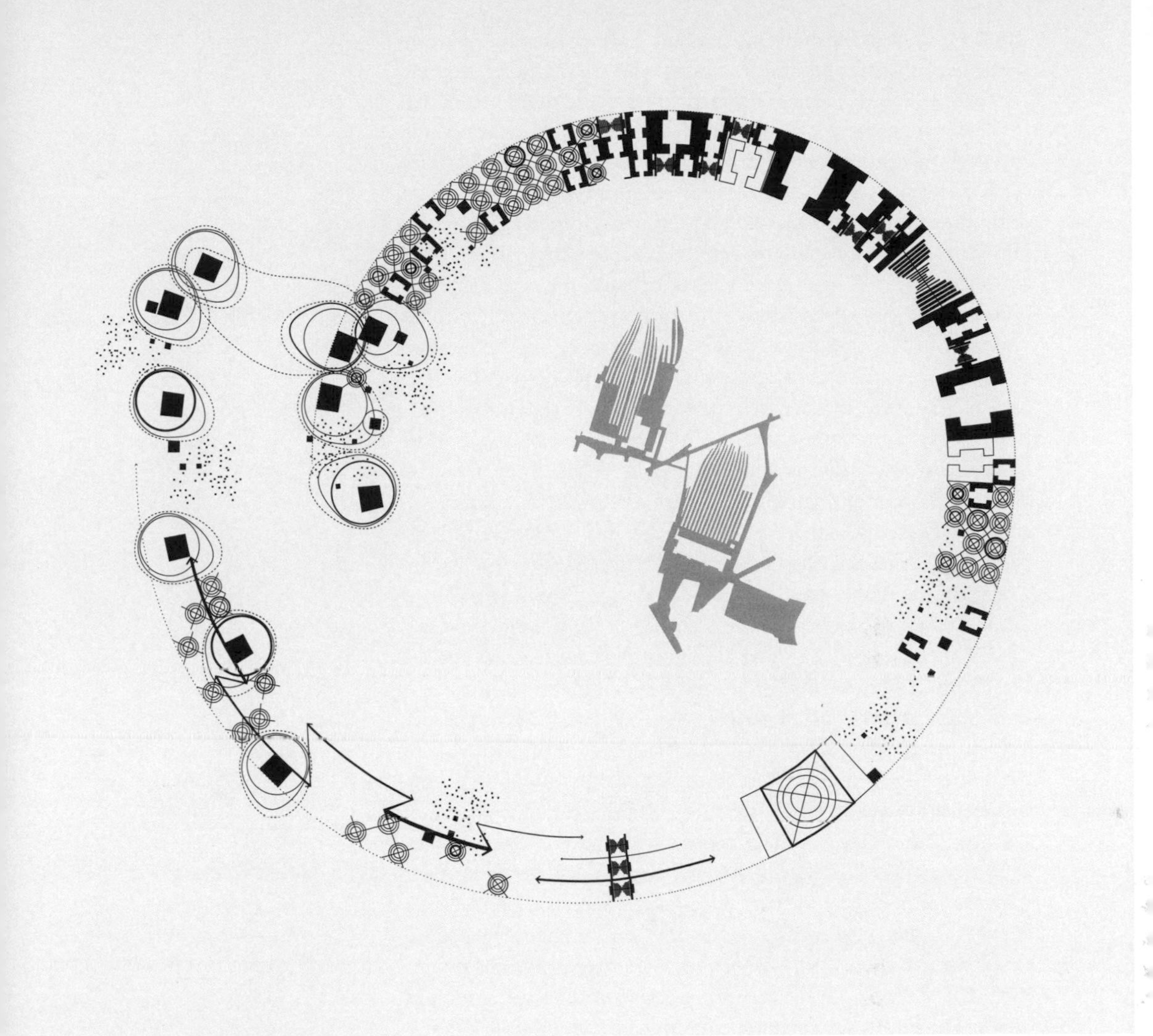

fig. 25 The score: from planning to the city with its potential times

impact (fig. 26). It also takes into account notions of mode of attendance and the number of entities present. Using these elementary categories, we grouped different places and moments of the city into typologies that generate map-scores based on their organization and the multiplicity of possible combinations. Each unit-point then develops individually between expansion, condensation, and retractability. To read a territory as space-time amounts to wanting to decipher a score with an incredible number of staffs, in order to take into account the different times of the city: the cyclical time of seasonality; the political time of the power of construction and destruction; the time of living things and growth cycles, which cannot be compressed; and project time, which is regulated and organized into phases.

Space-time stenograpy allows us to describe space and time simultaneously in order to navigate them. Our actions in the public domain, time and space, are multiple: we move, meet, browse, play, eat, demonstrate, rest, wait... These actions can also be summed up in an act of presence: being in the world. The symbols that represent them can develop their own forms according to their frequency, intensity, rhythm, and the variation of these actions. This spatiotemporal corpus, like a survival kit, contains only the most essential symbols (fig. 27). Experience will nourish and enrich this semiotic toolkit. The symbols are grouped into families that combine spatial footprint with duration and frequency. The family that most obviously combines time and space in this legend is standardized space-time: a soccer match, for example, requires a precise surface and its own time period. It is a practicable space-time governed by a common standard. Conversely, elastic space-time eludes all metrics. Other space-times are calibrated but not in a permanent way. Markets are spaces-times that appear and disappear repeatedly. They occupy a recurring period of time and a defined space. The public square is another interesting example. A fixed

fig. 26 Principle of notation based on the spatiotemporal footprints of the elements of the city

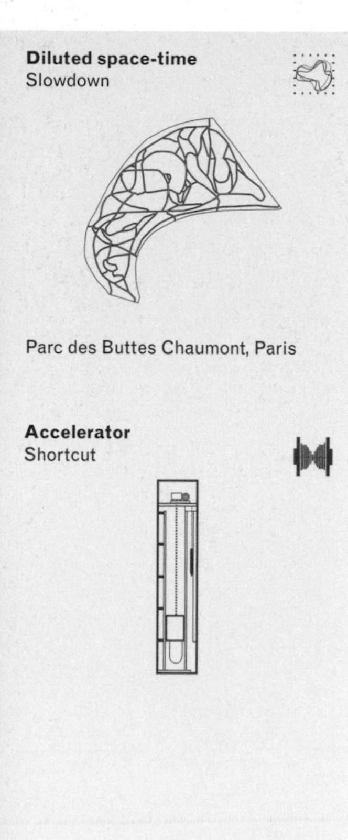

receptacle in terms of spatial footprint, it can nevertheless go beyond its limits, depending on how it's used. A shared place, it constantly redefines its boundaries of use and its time limits. There is also an almost-outside space-time, that of waiting. The immobility of the agents, their small spatial footprint, the distortion of perception that seems to extend this kind of time, and finally the fact that they are temporarily cut off from other flows—all this inscribes them with a parallel pattern. Finally, the notation reveals something that really is outside space-time: the death of a being, a collective, or an idea. By preserving it in one's calendar or in space (as a memorial, cemetery, monument, national holiday, and so on), the notion of duration is replaced by historical time. The lost physical presence is replaced by a fixed object or date.

Musical and choreographic notation make it possible to understand space-times. As for the structure of our dial, it is no longer a calendar or a linear score but a spiral, without beginning or end, winding up the timeline to integrate the cycle of the seasons into different urban units. The time of the city is translated by a system of interconnected spirals that makes it possible to read the simultaneous evolution of these spatiotemporal units. The dial is also a fractal structure that maps the relationship between space and time by adjusting this transcalar tool to the scale of needs, from buildings to neighborhoods. The "scope" that accompanies the structure of the spiral makes visible the micro-spatiotemporalities of a territory. This is the score in which the space-time will play out.

To test the model, we chose the city, a place with maximum layering of tempos, and more specifically the northeastern district of Paris (its 18th, 10th, and 19th arrondissements), one of whose main characteristics is the layering of rhythms: two train stations and their tumult, markets, residents, passersby, a canal,

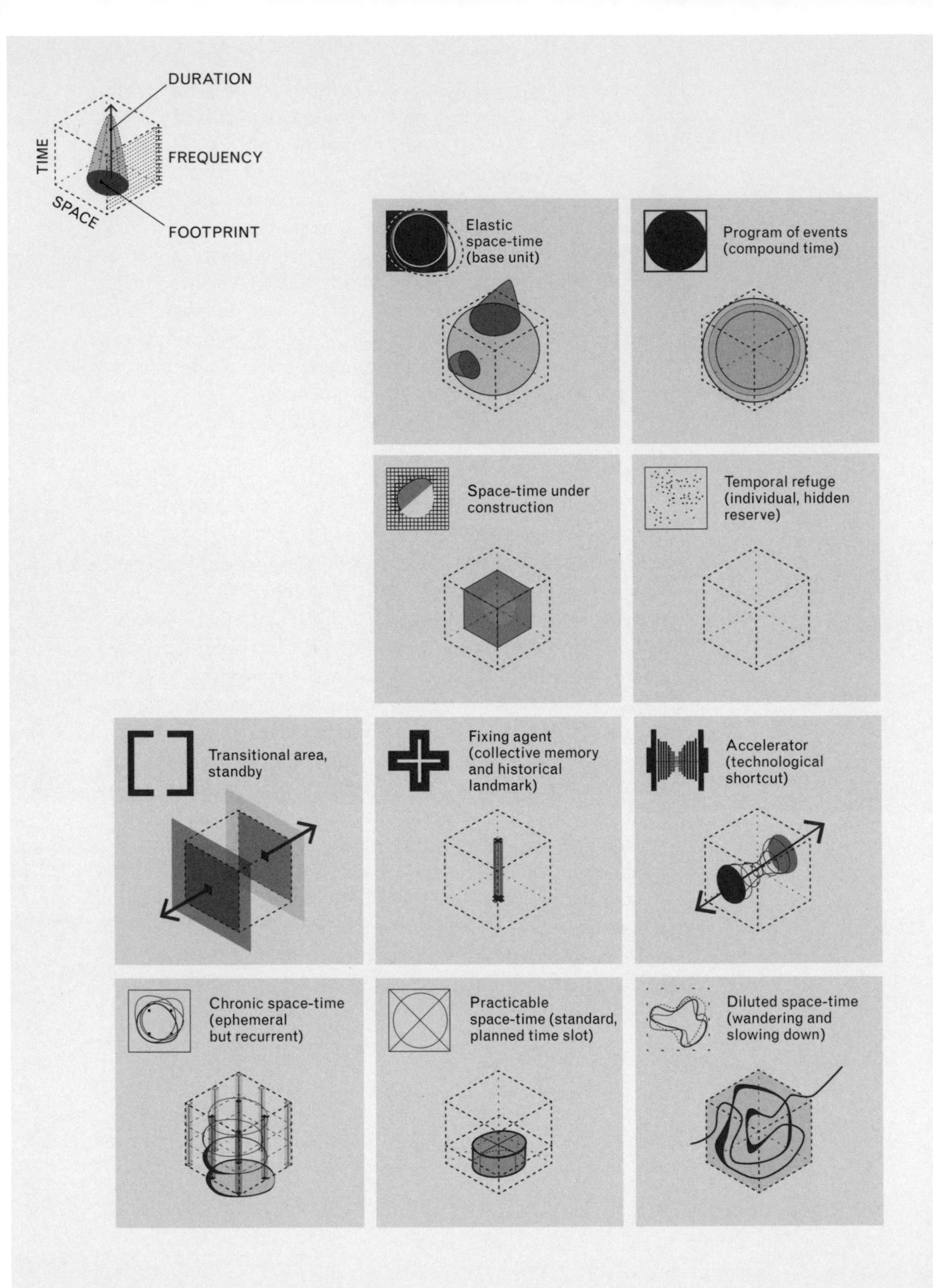

fig. 27 Stenography of space-time: drafting notes for the urban score

The stenography of space-time is a system of notation that allows space and time to be described simultaneously in order to navigate them or to develop projects there. The symbols are grouped into families that combine spatial footprint with duration and frequency. As a survival kit, this spatiotemporal corpus contains only the most essential symbols. Experience will augment this semiotic toolkit.

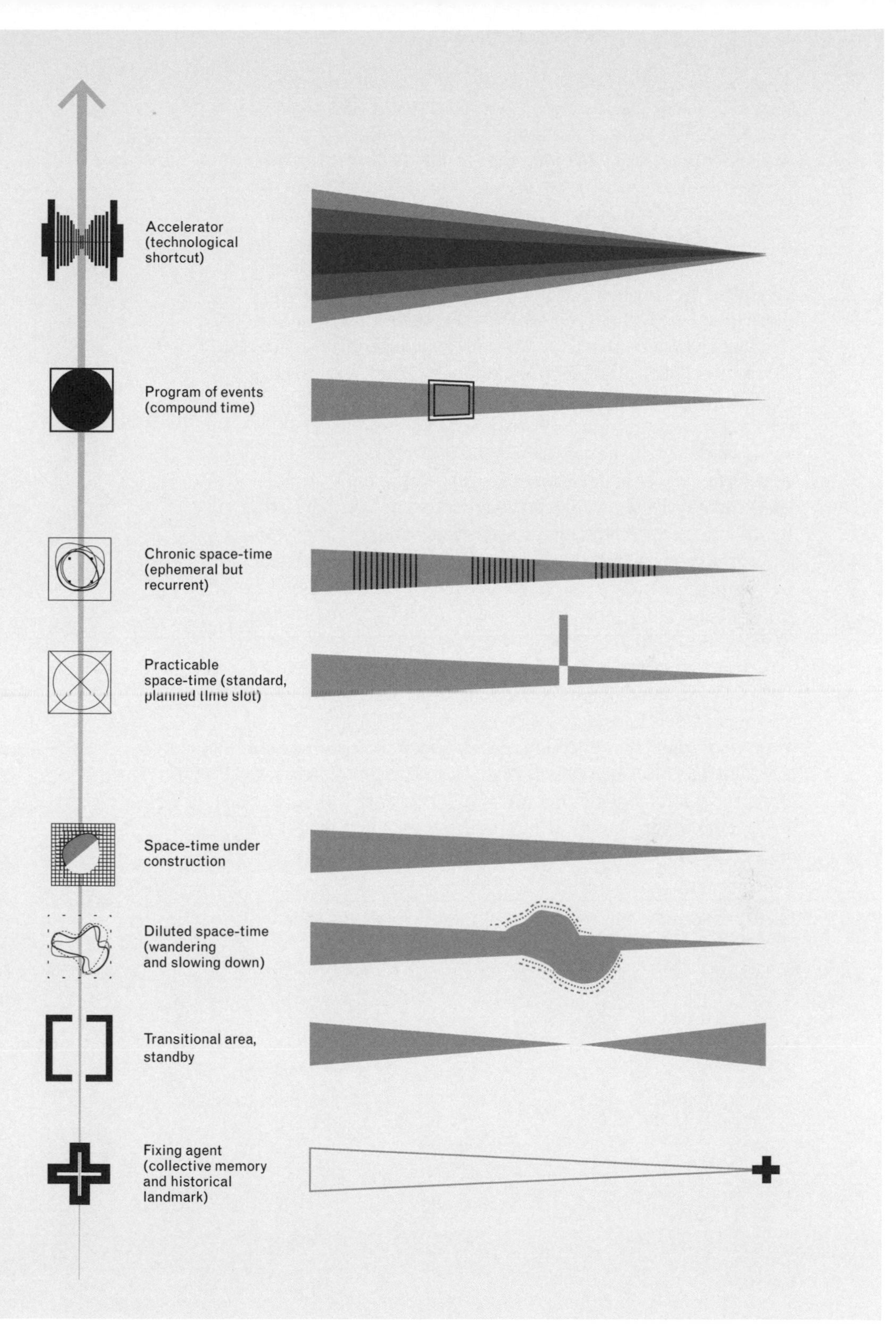
Accelerator
(technological
shortcut)
Program of events
(compound time)
Chronic space-time
(ephemeral but
recurrent)
Practicable
space-time (standard,
planned time slot)
Space-time under
construction
Diluted space-time
(wandering
and slowing down)
Transitional area,
standby
Fixing agent
(collective memory
and historical
landmark)

a park . . . (see map pages). Each urban unit in the district constitutes a coherent whole from the point of view of functionality and develops a score, as a cog in the global engine. This dial defines a climate that must compose music with the other scores of the city. Subsequently, the city can be seen as an archipelago of space-times in which certain islands merge, others overlap, and still others refuse any connection (see second map). The maps attempt to capture the neighborhood's rumors, moods, and climates. They make it possible to locate anomalies and other blockages: event bubbles, short circuits, clusters, accelerations, and instances of overlap that generate conflicts in places where rhythms collide to the point of cacophony. They also make it possible to identify irreducible cycles, repetitions, patterns, reprises, and themes, and they generate new questions: What are the instruments of the orchestra? Who is directing, and who is composing (fig. 28)? The history of musical notation teaches us that scores evolve depending on whether they are written down by the composer, conductor, or performer. Whoever takes control of the system of notation redefines its objectives. As Hartmut Rosa explains, "Knowing who sets the rhythm, duration, tempo, order of succession, and timing of events and activities is the arena where conflicts of interest and the struggle for power play out. Chronopolitics is therefore a central component of any form of sovereignty."[40] In the same way, the notation of space-times proposed here is a system of composition that allows polyphony, the reintroduction of the notion of cycle and season, a writing of simultaneity, and a multiscalar coordination of the different spatiotemporal cogs (fig. 29). The planner will be able to develop a polyphonic score, while the user will model an individual format by their movements and the force of the

---

40 Hartmut Rosa, *Accélération: Une critique sociale du temps* (Paris: La Découverte, 2010), 26.

1 A dial is linked to an urban unit and is part of a complex mechanism. It draws a temporal spiral presented on a site plan. This spiral can represent various temporal scales: a day, a year . . .

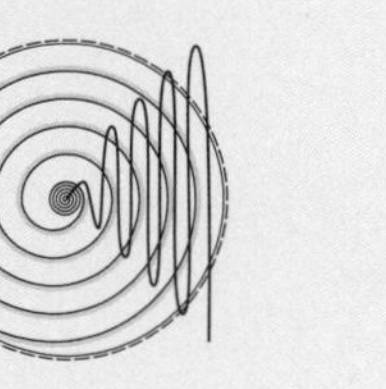

2 Bodies can enter and exit this dial. They are caught up in the movement and rhythm, integrated into this chain of space-times.

3 A dial is linked to other dials. It can be extracted for a diagnosis or project but it belongs to a rotating assembly.

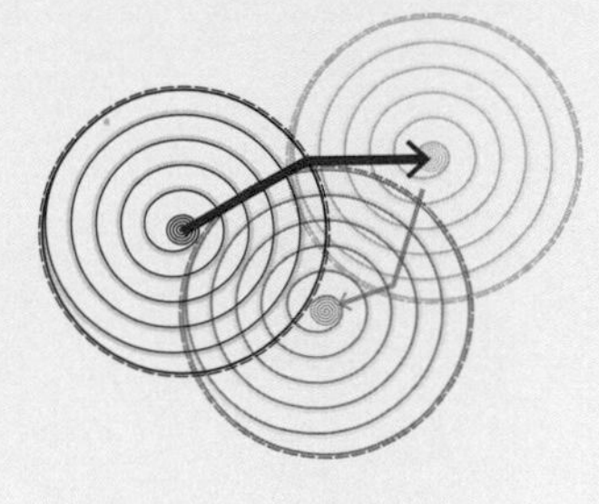

The dial allows the past, present, and future to be represented on the same drawing.

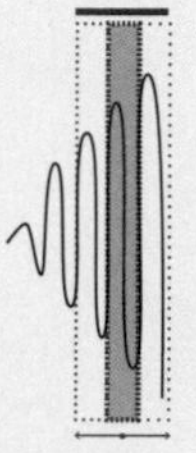

The dial can be a diagnostic and/or prospective tool.

4 What are the objects that link them? Space-buffer or springboard.

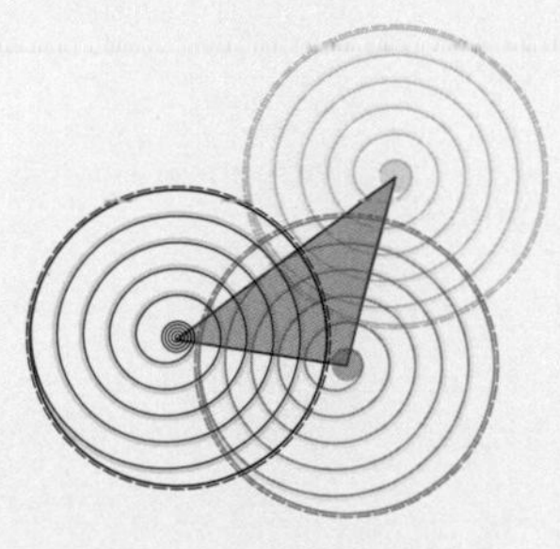

5 The selected urban unit determines the scale of the dial and of the new score.

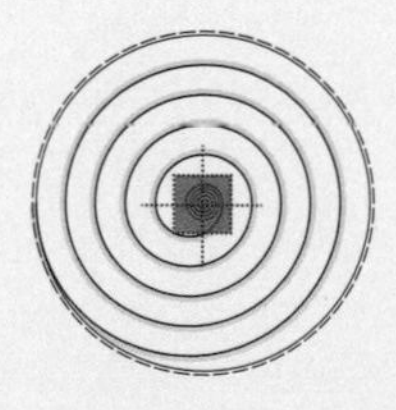

6 The arrangement of the dials arises from a subdivision of space-times and their linkage by places of connection. This division can vary according to the study's needs. Thus several dials can be grouped into a single one or, conversely, multiplied according to the scale at which the issues in question are to be understood.

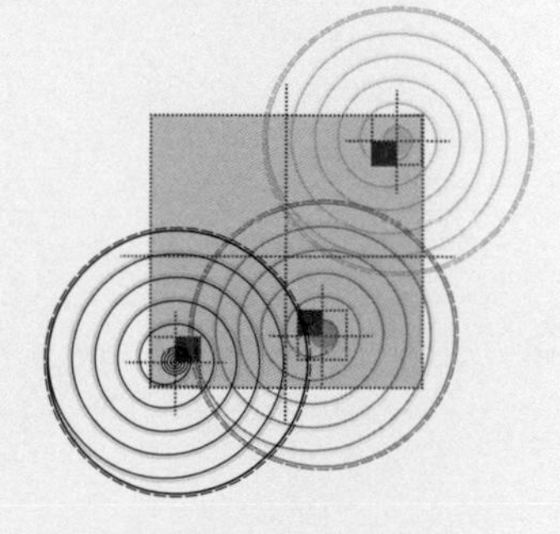

7 As with any mechanical apparatus, events (social movements, climate catastrophes, and so on) can block the dial, or even cause it to be bypassed.

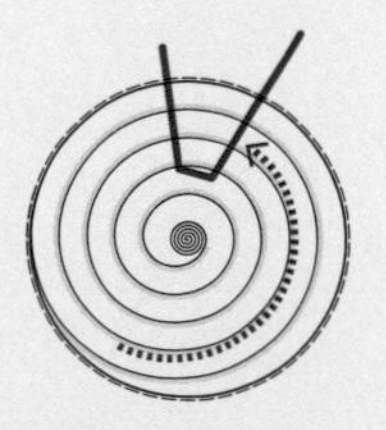

8 These blockages or other events can lead to a subdivision of the dials or a recomposition of space-time. Faced with the unexpected, points of life desert the shared dial in favor of their individual dials (space-time of the body).
An individualization takes place that can last a few seconds or a much longer time, at the risk of perpetuating the disharmony.

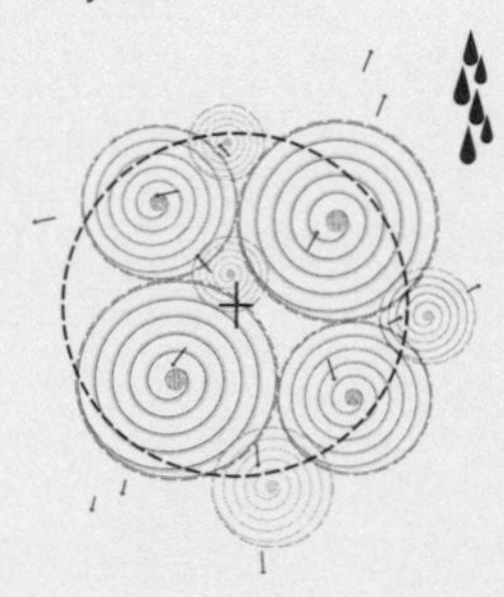

9 It is possible to extricate oneself from the dial. Examples include sites occupied by political activists and utopia.

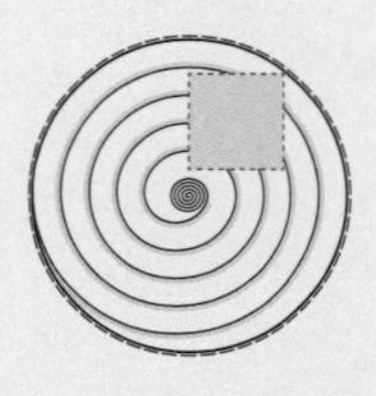

Fig. 28 Possibilities for the dial that describes a territory

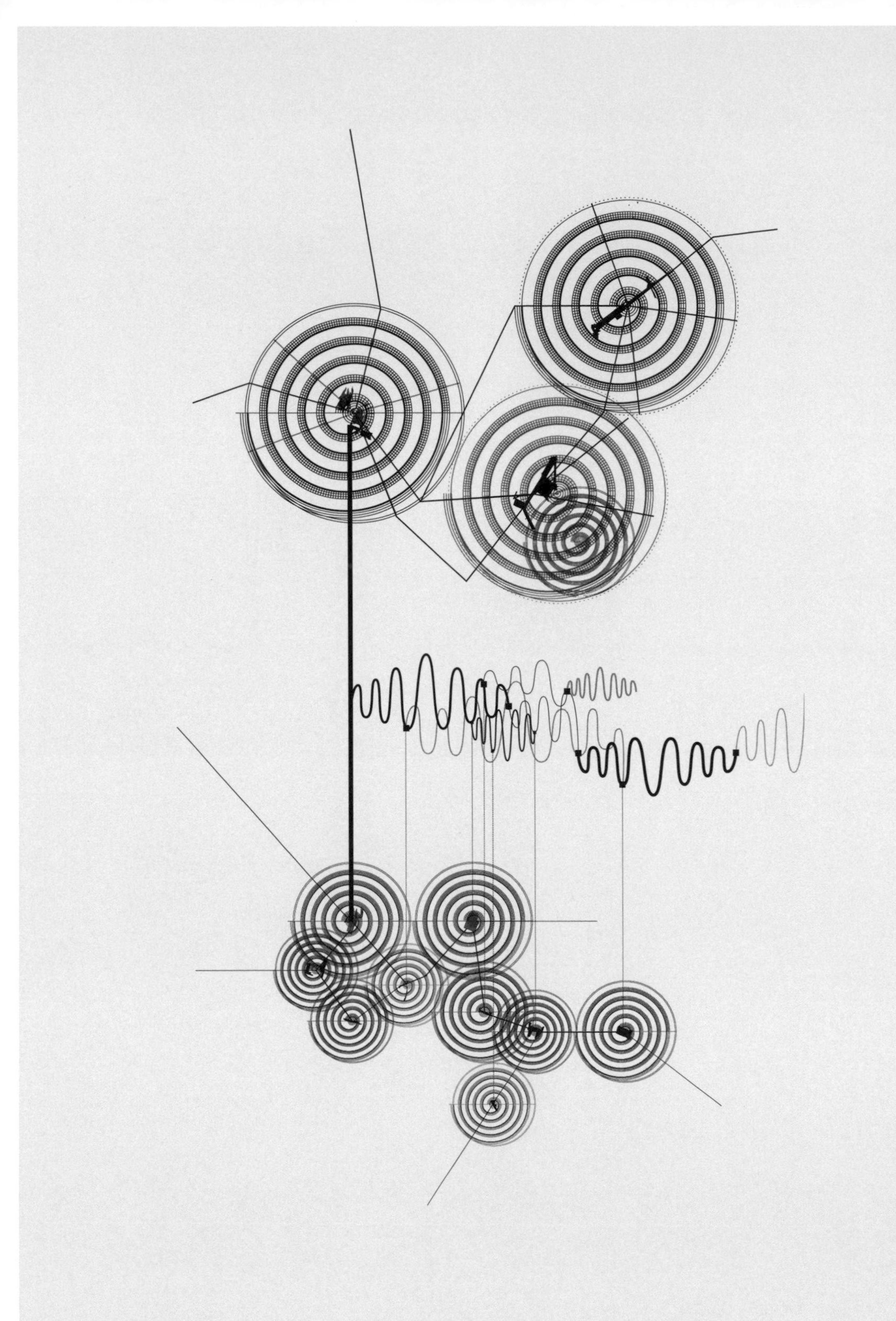

fig. 29 Depiction of the city

Desynchronized dials linked
by transit spaces

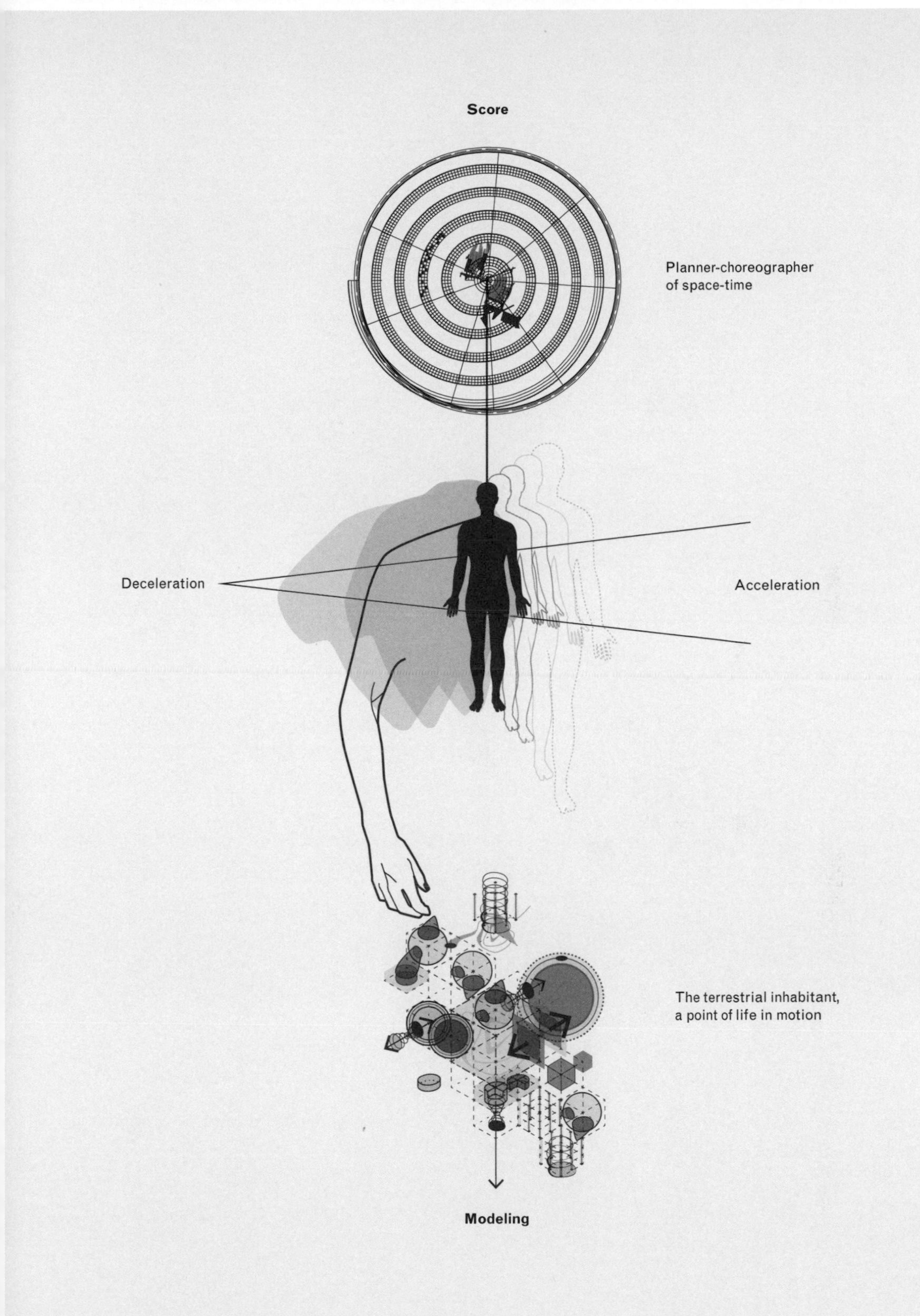

fig. 30 Relational tool between the planner and the user

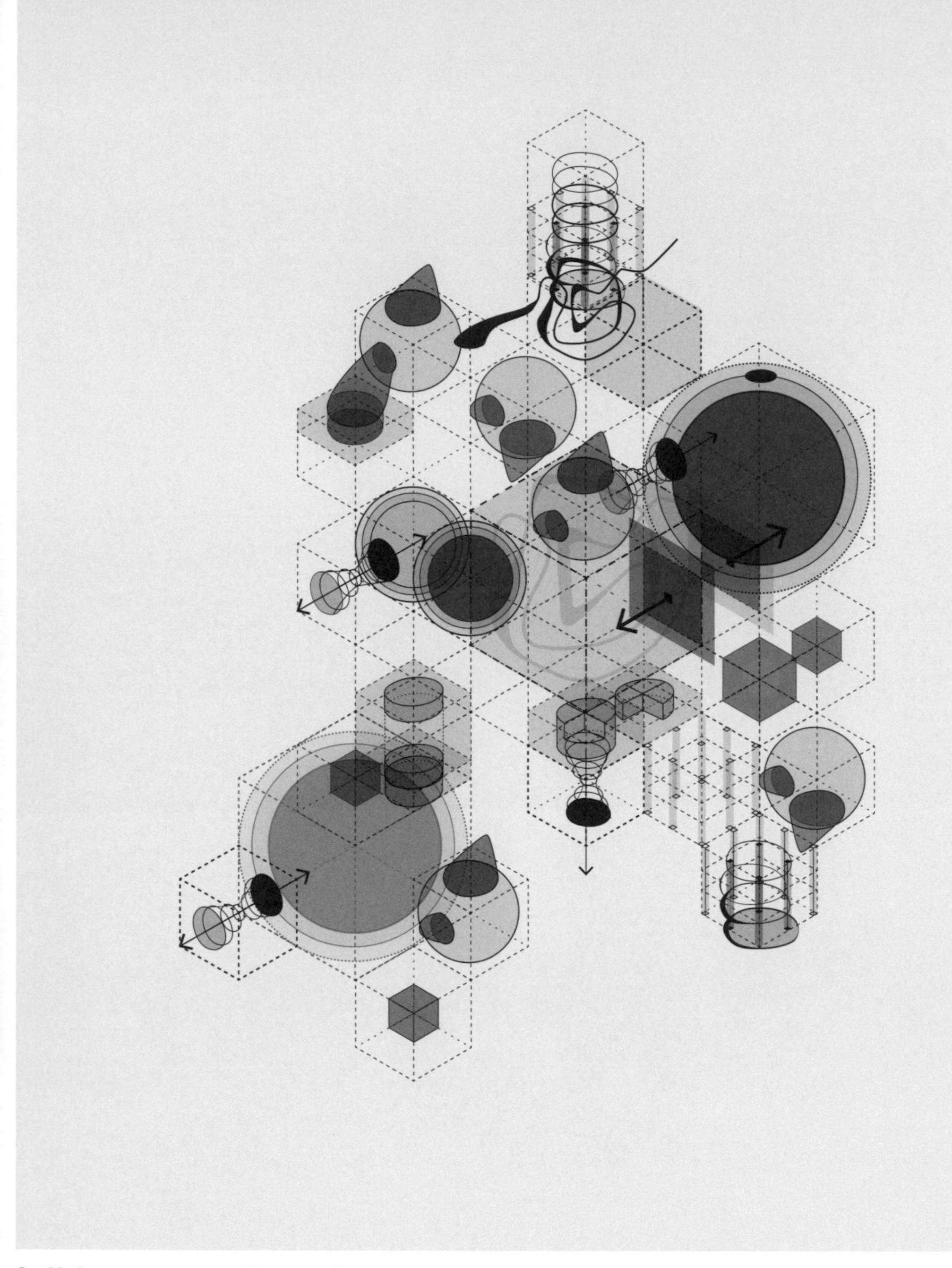

fig. 31 **Space: a monster of temporal architecture**

physical imprint of their body—an individual clock. The bodies in space-time also become the interpreters of the map, the sculptors of public space-time; as for the city, it becomes the expression of this public space-time constructed by the bodies and the movements that make it up. We are thus in a continual back-and-forth between the score (the point of view of the planner-choreographer) and modeling (the point of view and life of the user) (fig. 30).

It is striking how the map highlights the implicit norm that governs the organization of the city: a Western vision of the distribution of public time where little room is left for appropriation. It brings to mind a hyperdetermination of space—so much so that we can visually represent these spatiotemporal regions where the rhythm accelerates, where a specific dynamic develops that can snap up any point of life that crosses it, even if it means modifying that point of life's dial. But the map also shows the metamorphic character of shared time: humans and nonhumans cause spaces to move. Space-time is in motion. It remains, despite the efforts of the public authorities, a sort of monster of temporal architecture that is hard to control (fig. 31).

MODEL V
SPACE-TIME

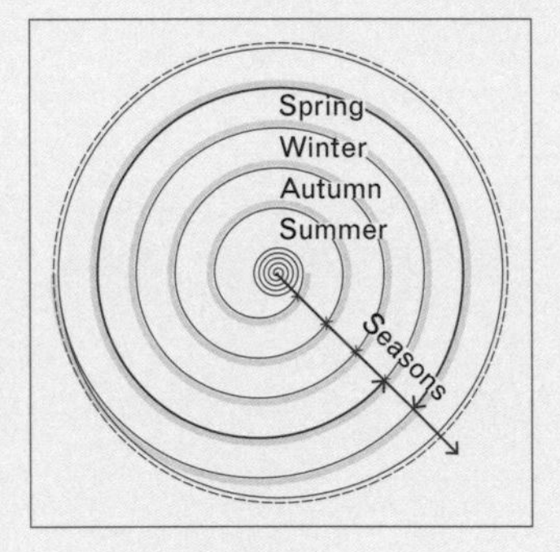

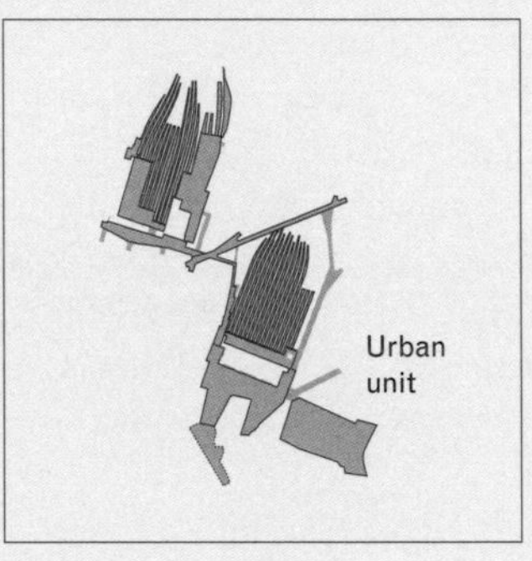

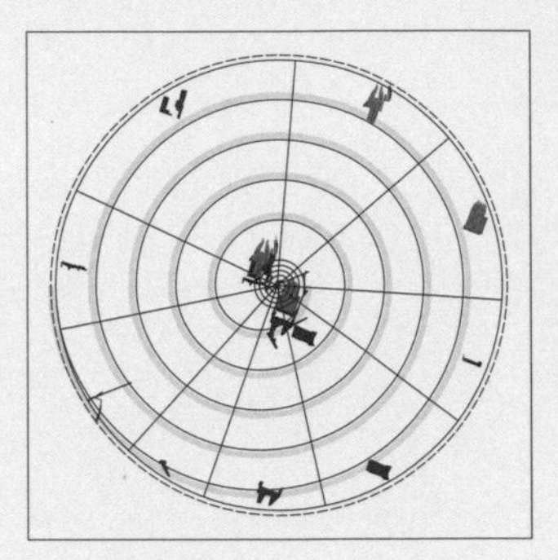

1

Generate the dial. Abandon the calendar's rectilinear structure and adopt the curved and spiraling structure of seasonality. This temporal structure is based on the regularity and cyclical nature of the seasons. The spiral has no beginning or end and stretches upward and downward in a dual motion that is both centrifugal and centripetal.

2

Identify the spatial unit whose space-time you wish to map. The coherence of the urban area determines the spatial unit. It works like a system, from the point of view of usage or spatial grammar. This unit of spatial coherence can be selected at different scales (from the district around the train station to the station itself).

3

Place this spatial unit at the center of your new dial. Layering them creates a spatiotemporal unit. The spiral of this spatiotemporal unit is then subdivided into pieces corresponding to its spatial subunits. Each of them is distributed around the edges of the dial, like new markers.

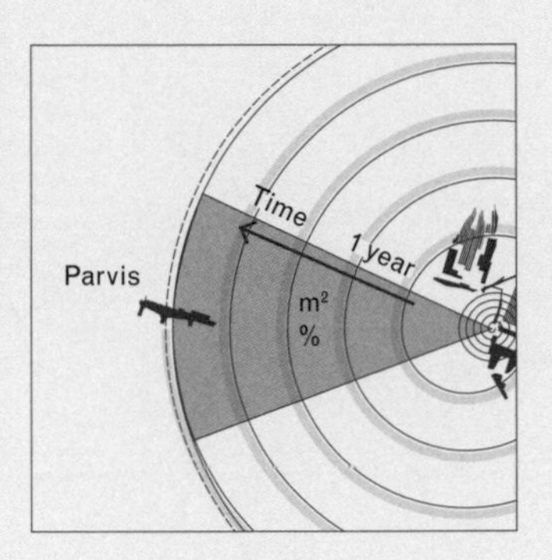

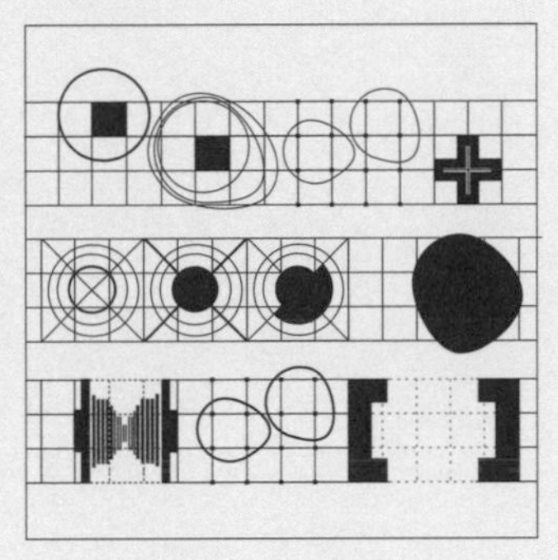

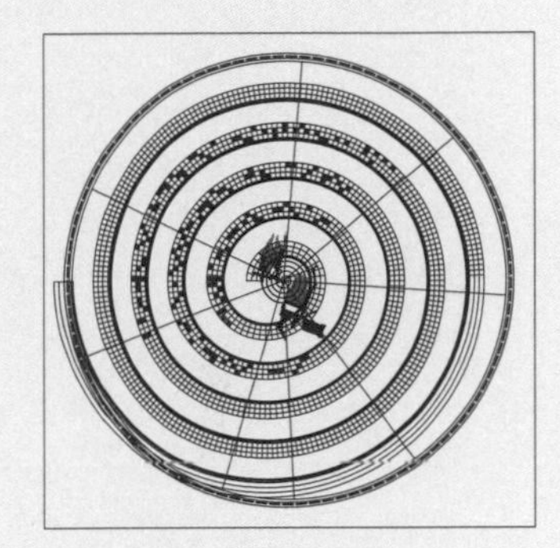

4

Create the score. The divisions are carried out according to the percentage each subspace occupies within the unit as a whole. Each "part of the dial" therefore corresponds to a spatial subunit over a period of time of about one year, or four seasons (which means that the spiral of the seasons crosses all the dials and links them "climatically").

5

Create a legend based on the qualities identified in the urban space-times. The legend contains a new system of notation to describe what characterizes these space-times. A musical scale has been developed as a basis and can be expanded or corrected based on experience. The legend must take the perception of time into account, as well as modifications to the plasticity of spaces produced by variations in how the space is used: waiting, sudden acceleration, disintegration, time in suspension, the stability or elasticity of moments, and so on.

6

Subdivide the timeline of the spiral into micro-spatiotemporalities. To do this, build a grid that is molded to the shape of the spiral. The boxes are the structure of the score, the staves in which this space-time will play out. Next, place the micro-spatiotemporalities (from the legend) into the boxes/scores corresponding to the space (part of the dial) and the season (line of the spiral) being studied.

MODEL V
SPACE-TIME

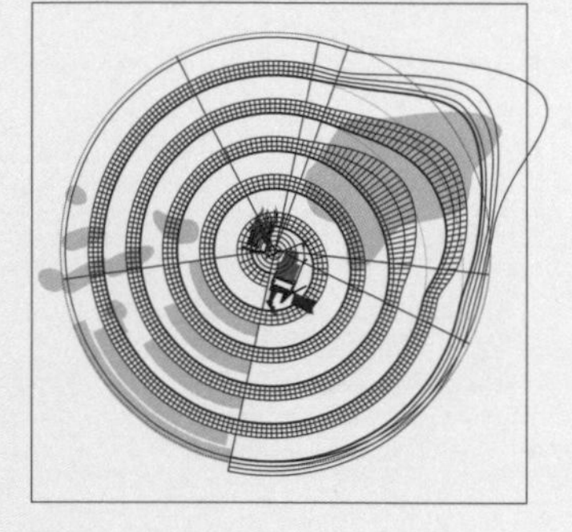

7
Once all the boxes are filled by micro-spatiotemporalities, a landscape emerges. This is the landscape generated by the space-time of the studied unit. Variations in the landscape can be sorted into large visual groupings. The interpretation phase of the score will allow us to contemplate its subsequent manipulation: depending on the project or the choices of the inhabitants, the grid or score may be molded to accommodate new demands for the use of space and time. Space-time is socially sculpted. But it is also potentially molded by natural variations (water, vegetation, climate) that disrupt spatiotemporal uses of resources (see second map). Finally, given that a dial reflects a space at a given time, it does not constitute the fixed and timeless data of a spatial situation. Thus more dials must be added to offer a correct spatiotemporal vision of a territory.

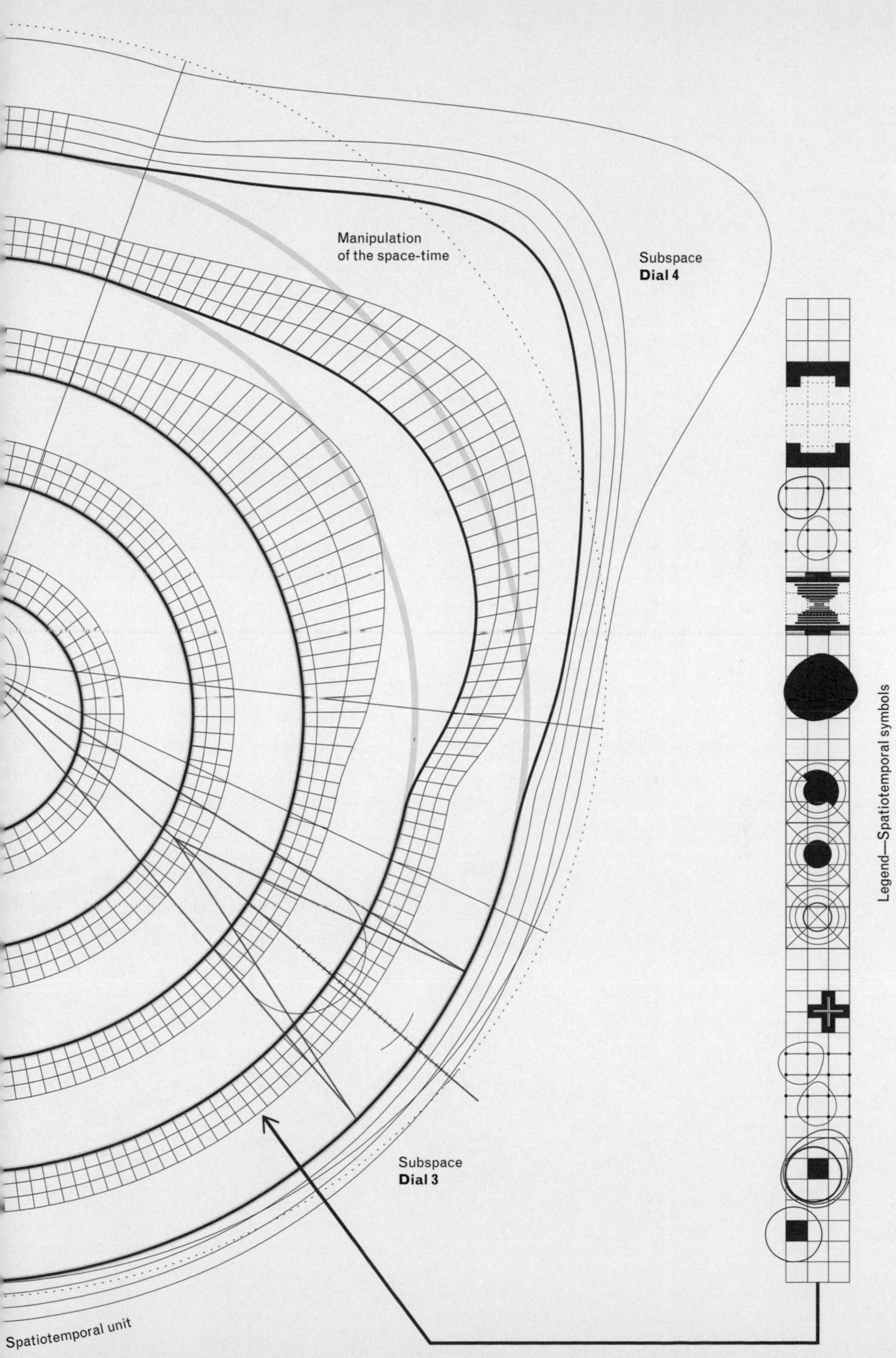

Manipulation
of the space-time
Subspace
Dial 4
Legend—Spatiotemporal symbols
Subspace
Dial 3
Spatiotemporal unit

Stenography of space-times

Elastic space-time (base unit)

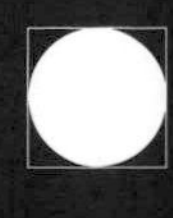

Program of events (compound time))

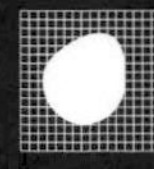

Space-time under construction

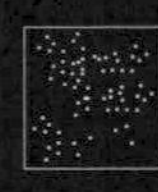

Temporal refuge (individual, hidden reserve)

Transitional area, standby

Accelerator (technological shortcut)

Chronic space-time (ephemeral but recurrent)

Practicable space-time (standard, planned time slot)

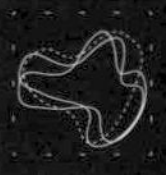

Diluted space-time (wandering and slowing down)

Fixing agent (collective memory and historical landmark)

MAP V SPACE-TIME

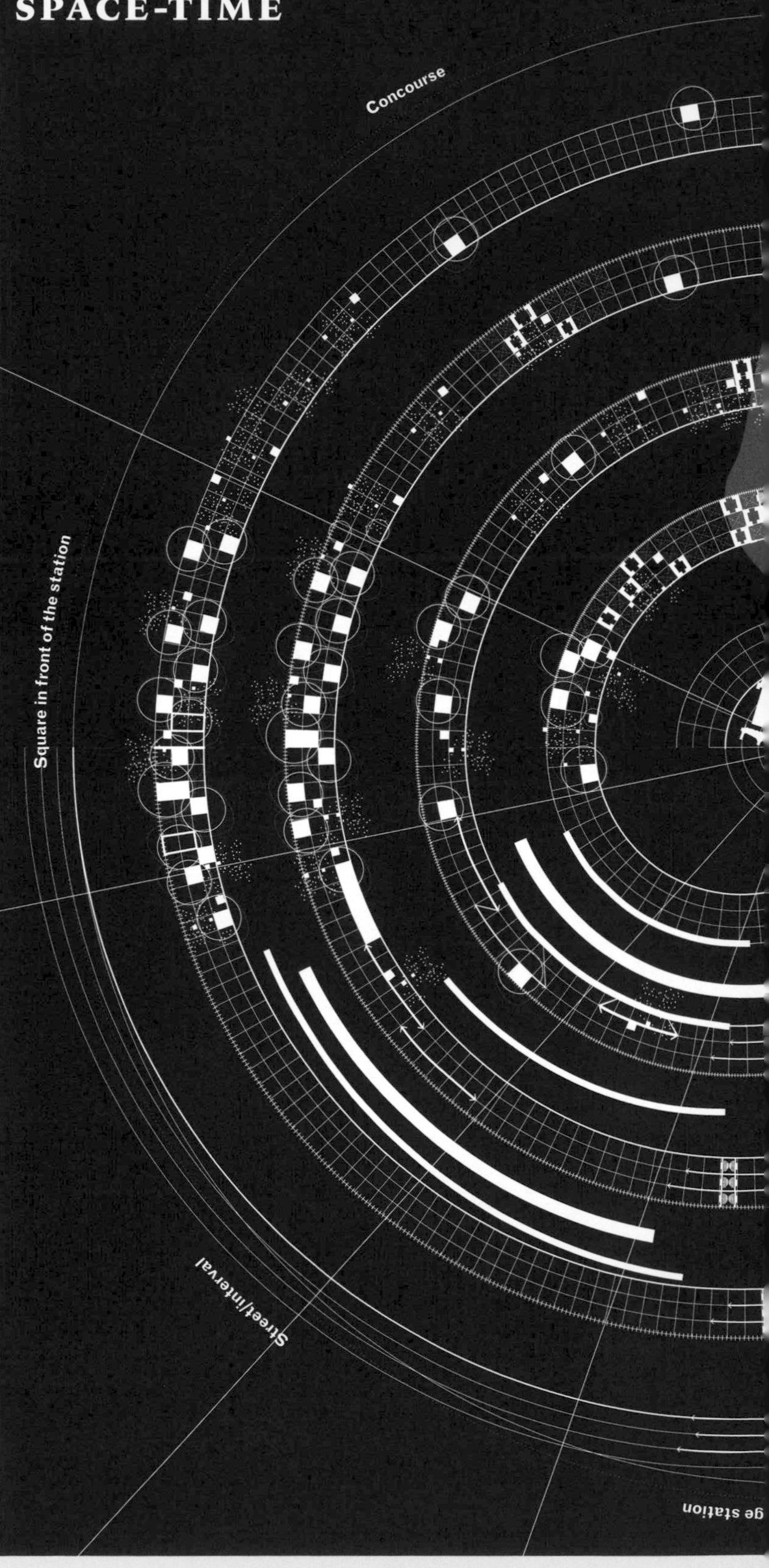

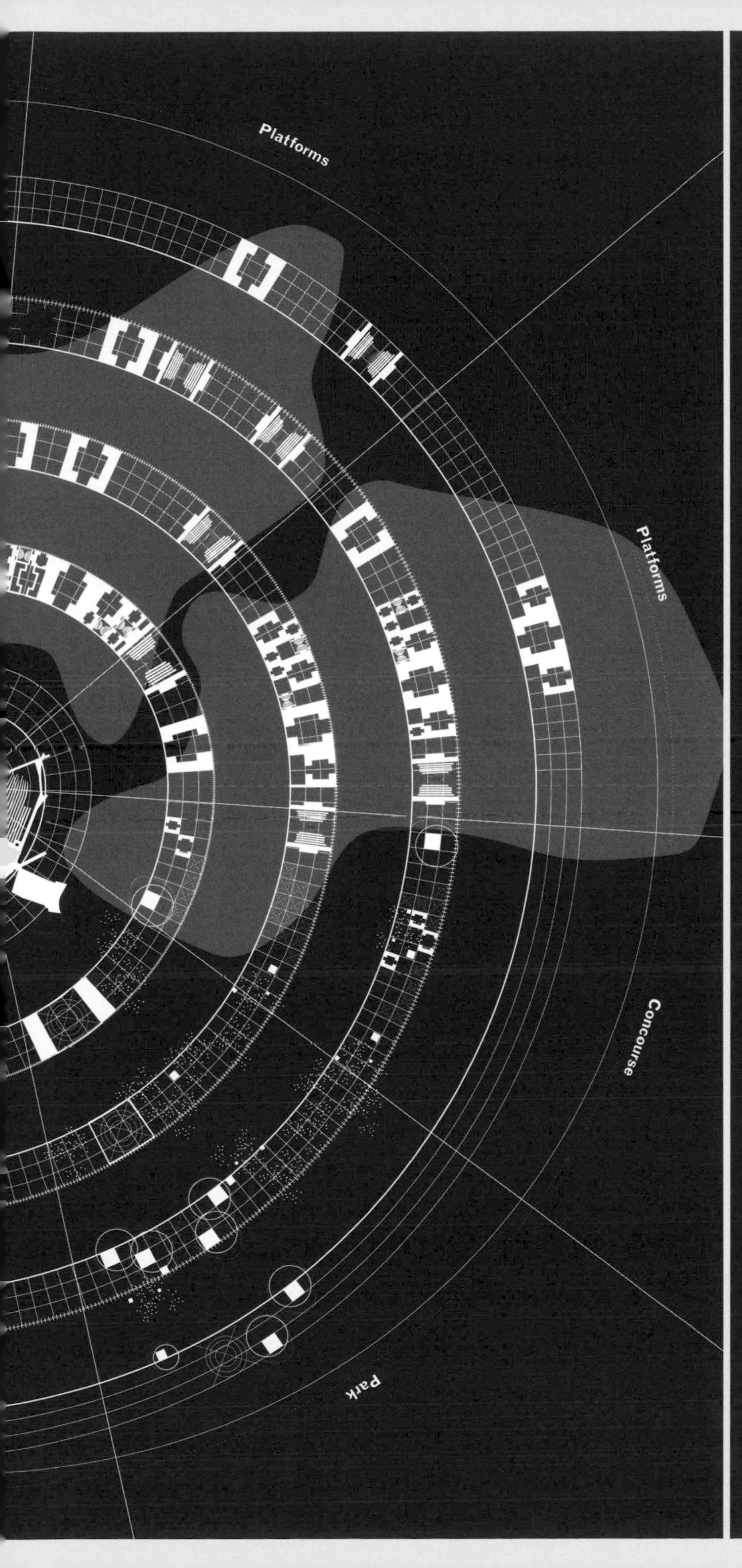

Space-time of the squares in front of the stations

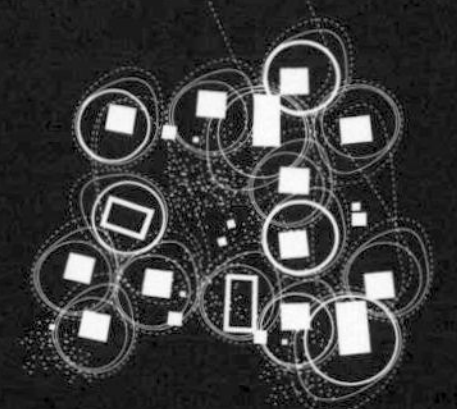

Planner-composer: a smooth space that is a receptacle for an urban polyphony with variable geometry.

User-performer: a pivotal point between city time and travel time, during which individual dials contort and rearrange on contact with each other, creating a dynamic agglomerate.

Space-time of streets connected to the bipolarity of train stations

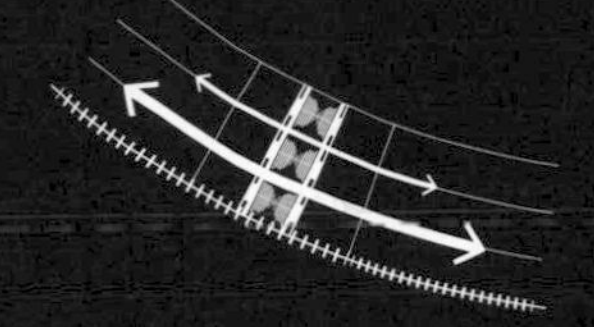

Planner-composer: a patched transit conduit to compress time and increase the efficiency of the linkage.

User-interpreter: an attempt to connect that is disrupted by the flows of the city that pass through it and which make it to adopt various postures of avoidance.

Space-time of the platforms

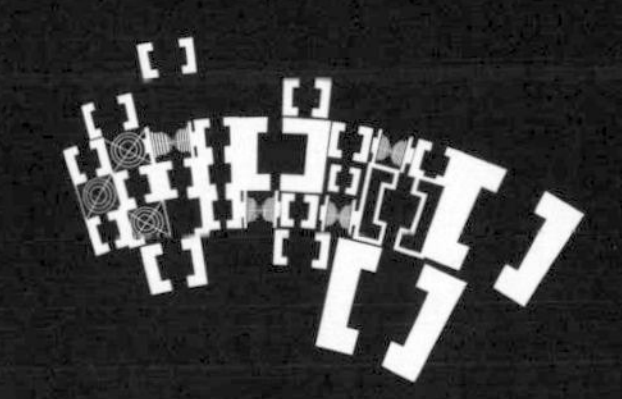

Planner-composer: a stationary line for boarding or disembarking from a collective accelerator.

User-interpreter: a marginal and heterogeneous space where slowing down gives way to waiting.

A transitional space of decompression toward a shortcut that brings geographies together.

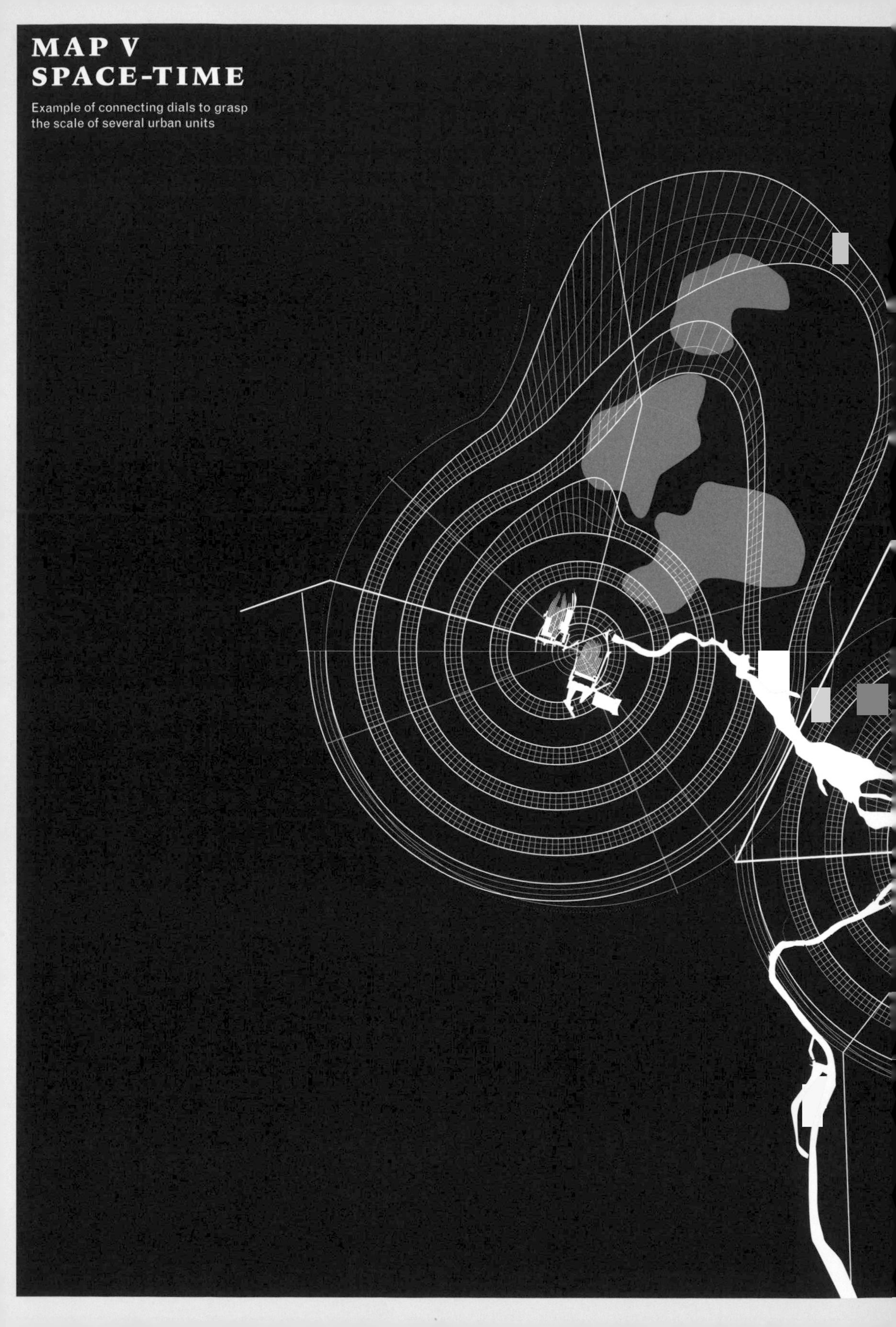
MAP V
SPACE-TIME
Example of connecting dials to grasp
the scale of several urban units

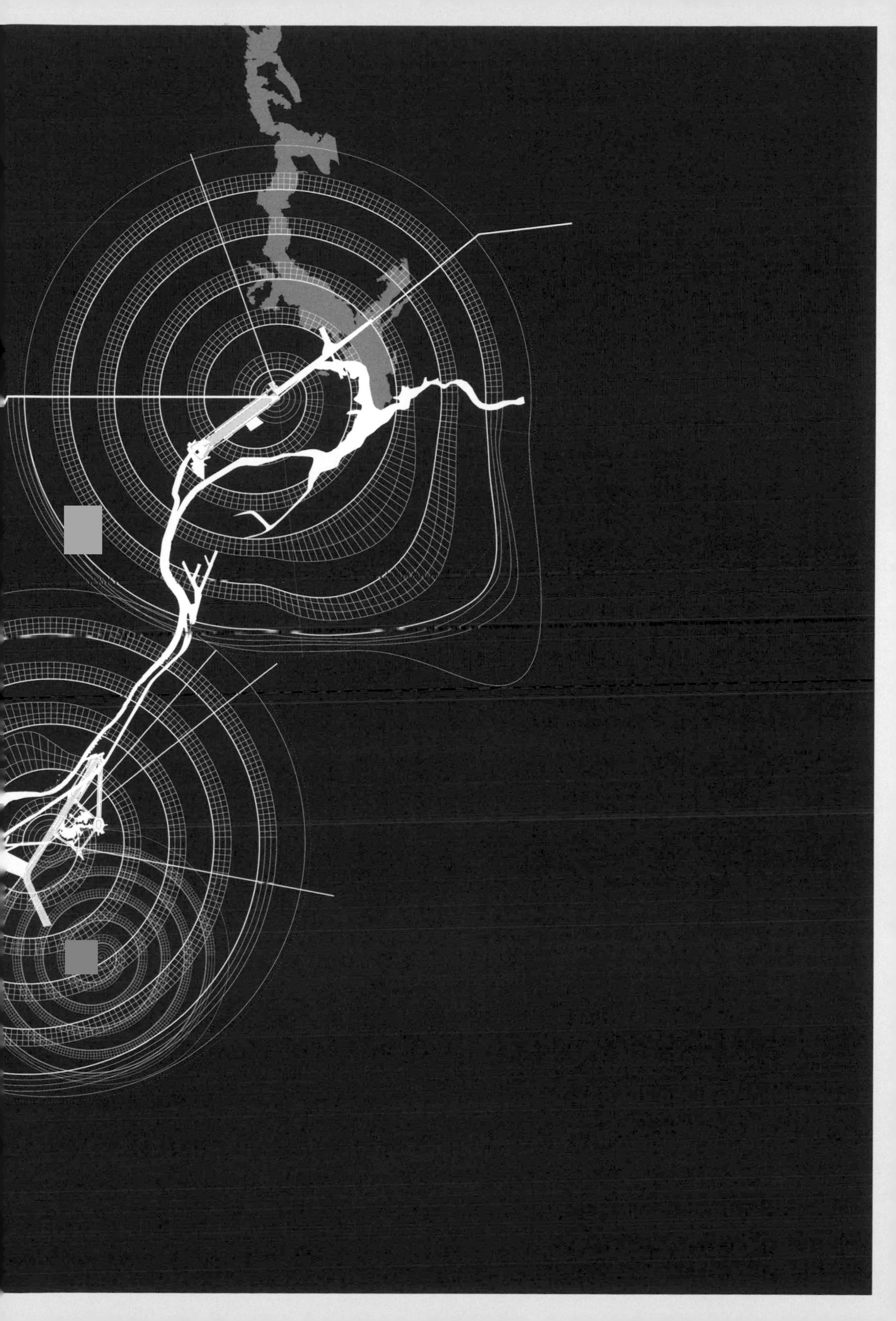

TRANSFORMATION V
SPACE-TIME

Linear time . . .

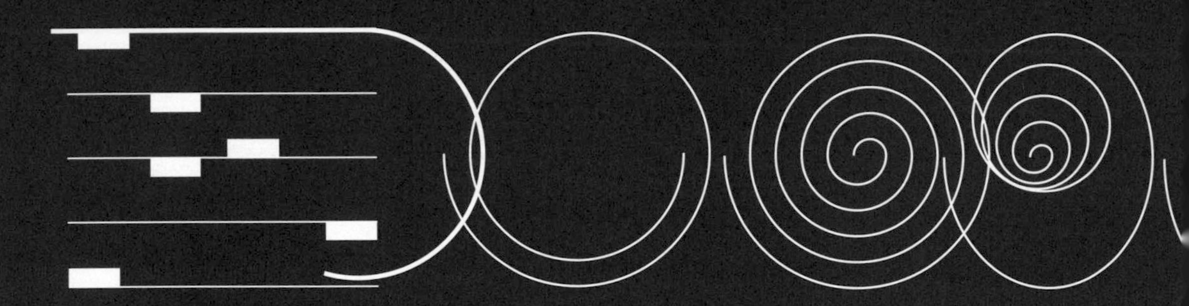

. . . spiral of space-times.

fig. 32 **Holes in a territory (Marnes/Aube)**

MODEL VI

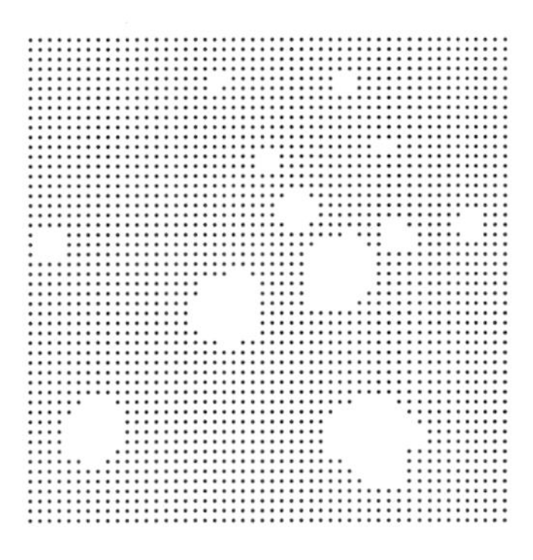

(RE)SOURCES

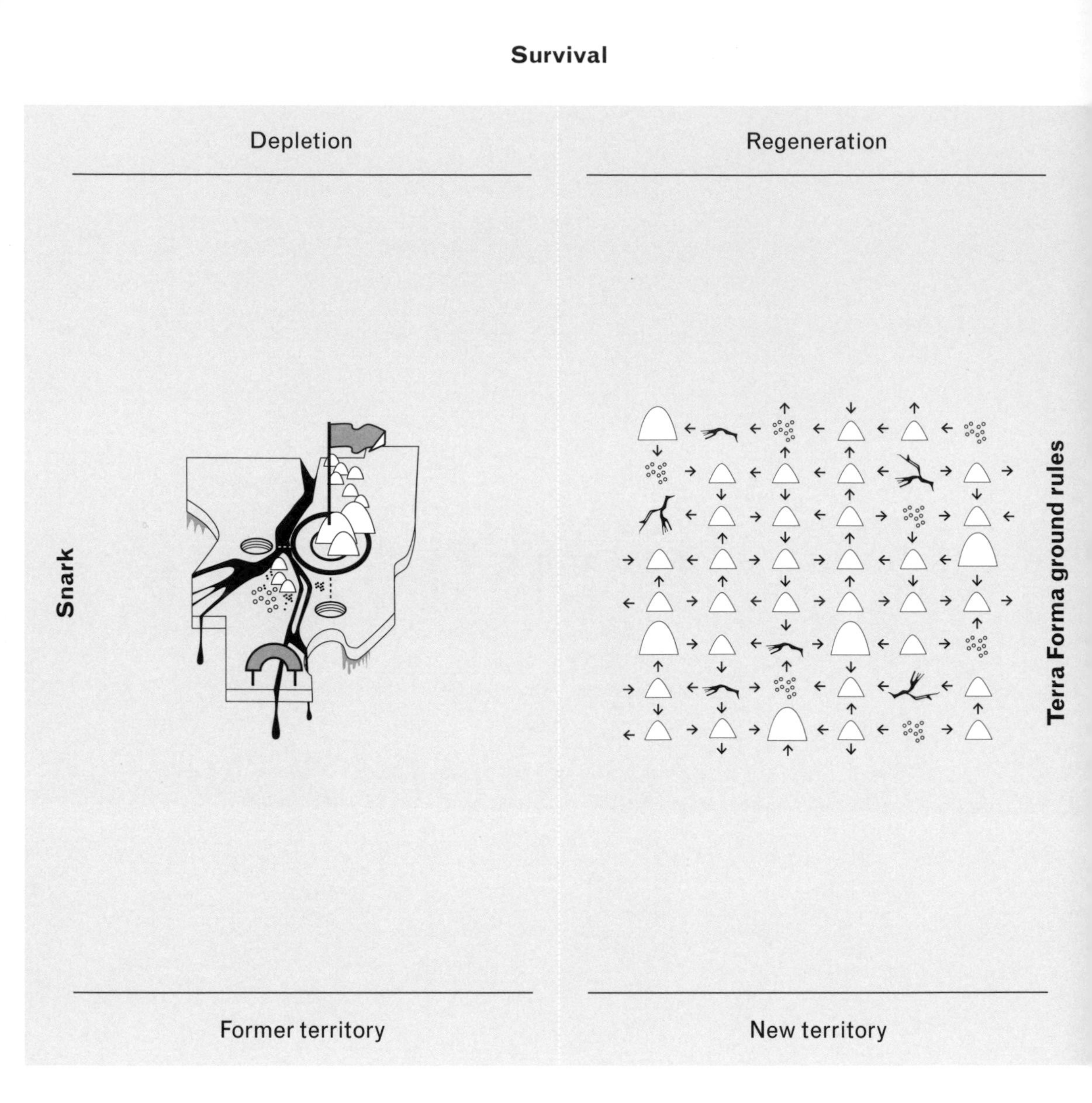
Survival
Depletion
Regeneration
Snark
Terra Forma ground rules
Former territory
New territory

(RE)SOURCES

Equipped with our first models, we have surveyed a profuse and living Earth. By adopting the point of life and point of view of the animate entities who traverse it, we have observed their ways of shaping the ground, of composing and constantly modifying landscapes and borders. We might even say that there are no bodies in the world, only bodies producing overlapping world-spaces, altering each other, drawing energy from each other—animals breathe the breath of plants, while plants inhale our breath in an act of co-breathing. In this space, defined by the unique ways that bodies are put into action rather than by a system of coordinates, how should we orient ourselves, act, or make a map? By what new means should maps be constituted? Based on this new point of reference—living things—the second part of our inquiry will investigate how to translate the maps into political terms to help us make land development decisions. We aim to unravel the consequences of this paradigm shift and contemplate ways to make territories that integrate affects and bodies—and, above all, to understand how the intertwining of living things and soils implies rethinking how we use the Earth.

One prevailing use treats the Earth as a resource.[41] It's a well-known principle. Let us recall its basic ideas, as always, by means of a drawing. The exploitation of a territory's resources occurs over several stages: the identification, isolation, and encapsulation of resources, then the establishment of infrastructures to extract and collect the animate or inanimate matter to be exploited—water, forests, minerals, and so on (fig. 32). The model represents this process using "suction cups." The sites

41 Recent studies have highlighted the long history of this extractive relationship to the Earth. See, e.g., Phillip John Usher, *Exterranean: Extraction in the Humanist Anthropocene* (New York: Fordham University Press, 2019).

where resources are concentrated are isolated under a bell jar to maximize their productivity (fig. 33). Industrial areas, areas of intensive agriculture, forestry, fishing grounds, tourist parks, mines, and quarries form "bubbles" in the territory. Suction is applied until their existing energy is depleted. The means employed to extract resources from these areas impose a different mode of operation from that of the rest of the territory, modifying the composition of the soil, the climate, and the living conditions of animal and plant species. These suction cups are not inert; they re-create what we call "biomes." A biome is a biotic unit, an ecosystem; with industrialization and the extraction of resources, these biomes are no longer natural but rather have been made artificial. They become "transnatural biomes" that re-create a closed world governed by strict standards—for example, in incineration plants or nuclear power plants. The proposed model includes representations of these new biomes with their zones of annexed life, sometimes including very distant elements. It borrows its visual structure from the Petri dish, the round plastic dish used in laboratories to isolate a substance, cultivate it, and make it usable in various experiments. Each dish encapsulates resources, so that the annexed area is isolated from its vital context, transforming the ecological conditions of the site. When the resource is depleted, the suction cup is removed, leaving a hole, void, wasteland, or ruin. Abandoned commercial zones, mono-oriented infrastructures in decline, and depleted mining areas generate enclaves, as well as a halo of dead spaces surrounding formerly exploited territory. A zone of phantom spaces also unfolds around the suction cup: invisible spaces that are not included in the balance of spaces devoted to resources, but which are nevertheless necessary for the suction cup to work. They often double its surface area. Access to resources, their extraction, and their use thus shape territories.

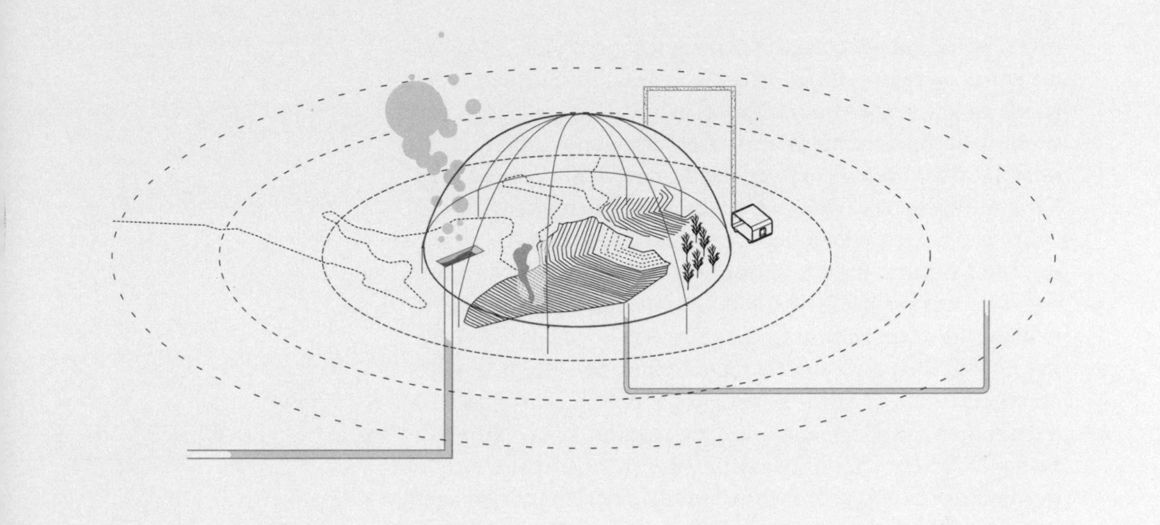

fig. 33 A typical suction cup

A territory is isolated and placed under a bell jar in order to exploit its resources

The first map is a mosaic of several territories in which each of the industrial biomes is representative of a mode of exploitation. This map therefore depicts typical functions of territories rather than the territories identified (fig. 34).

Intensive exploitation leads to the following cycle: appropriation of the territory > suction cup on natural resources > holes, ruins, and abandoned territory > rehabilitation. This model is the basis of the principle of renewal of "city within the city" sites, rehabilitation projects for wastelands, remediation of polluted sites, and so on. Ecology and economics are based on this model, which generates work for developers and investors but is depleting the land at a breakneck pace. Of course, relationships to resources take other forms and are not limited to intensive exploitation, especially in recent years. Other models have been developed, such as reconversion processes, ecosystem services, and the creation of nature sanctuaries.[42] The concept of "ecosystem services" implies maintaining the existing status through a system of ecological compensation. But this system, based on the notion of circularity, remains faithful to a notion of resources as services rendered to humans by nonhuman worlds. In this kind of practice, resources are isolated from the rest of the territory: the suction cup system opportunistically captures limited resources; protection isolates them. Whether suction cup or bell

---

42 The notion of "ecosystem services" is sometimes used to confer economic value, thereby creating a place for nature in development projects. The argument is as follows: planting trees is expensive, but it is a profitable operation in the long term (it increases the value of land, decreases health costs, and so on). This "landscape capital" often enters the economic balance of real estate development. But this notion of service rendered by nature is criticized today as utilitarian. Geneviève Azam makes a connection between our atomized conception of living things and Taylorism: the notion of "'environmental goods and services' has not fundamentally modified the model and has reinforced an anthropocentric, utilitarian vision of nature. Ecological balance is thought to depend on economic efficiency attained through markets for these goods and services, which, by causing prices to 'emerge,' encourage behavioral change and ensure optimal management of these environmental goods and services." Geneviève Azam, "Réduire le vivant pour le fabriquer ?" in *Les Limites du vivant*, ed. Roberto Barbanti and Lorraine Verner (Bellevaux: éditions Dehors, 2016), 367.

jar, it is more or less the same conception of "uses of the Earth" at work—the creation of a closed system that ignores any intermediate areas, those unconsidered parts of the territory.

If we want to better understand the type of relationships between humans and the Earth involved in each case describe above, it would be useful to take the detour through relations we are more familiar with: those that govern human working relationships. The exploitation of resources without compensation and to the point of exhaustion is strikingly similar to slavery. Ecosystem services fall under a form of wage labor, with all its well-known limitations: a type of contract established on the basis of a principle of exchange, but where equality is often feigned and violence hidden. The processes of reconversion, remediation, and rehabilitation belong to a form of compensation and care granted after services rendered, as in the case of a pension or retirement. As for the creation of sanctuaries, it perhaps confirms, in a pessimistic way, the impossibility of any relationship without exploitation. This reading of relations tries to illustrate the current ecology of resources between intensive exploitation, the desire for ecosystem services, the need for regeneration, and the temptation to turn the wild into a sanctuary.[43]

As architects, artists, and researchers, we inherit these sites and are asked to regenerate, rehabilitate, and repair them. But how can a depleted territory be repaired? Some would say that we have to get to the heart of the problem and put an end to exploitation. But reality being what it is, we must also find solutions for these sites, which are now part of our territories and our landscapes. This means formulating new recipes, starting from the observation that our current ones are biased by a philosophy of conquest and exploitation, with services drawn from an inert and available nature, and land taken and zoned, reflecting an

43 Here we observe a striking parallelism between economics and ecology, both united by the same conception of the exchange of goods and natural goods as part of the *oikos*, the household. If the world around us (property, land, natural beings, even "subordinate" beings like women, children, and animals) is a house, it is up to the master of the house to know how to manage it well. If it is not a house, we need to fundamentally rethink our conception of economics and ecology.

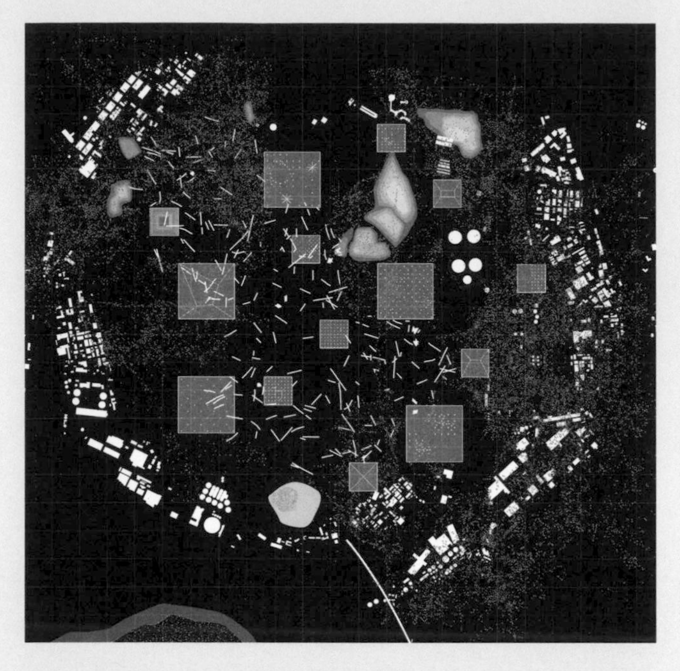

Suction cup 1
An exploited forest, typical of the temperate forests of France. It has been thinned out. The cut wood becomes material that undergoes a process of transformation and conditioning, from felling to its reuse in the form of lumber or fuel. A system of regular parcels of land is put into place so that the cutting process can be managed efficiently. After the marked trees are felled, the next step is transporting them from the site, which involves cutting branches and bringing logs to a passable road or waterway. The logs lying on the ground will not stay there for long; they will be transported to nearby sawmills or processing plants. This is why it is so important that the "machine-forest" be located close to infrastructure. A logging truck comes to collect the wood and deposit it in a sawmill's log yard. There, the wood will be cut, dried, and treated, then distributed to different places. Finally, it is time to replant and wait. This forest-in-the-making will have 30 to 50 years (the turnaround time for trees to regrow in an exploited forest) to try to store surplus carbon, before it is cleared out again.

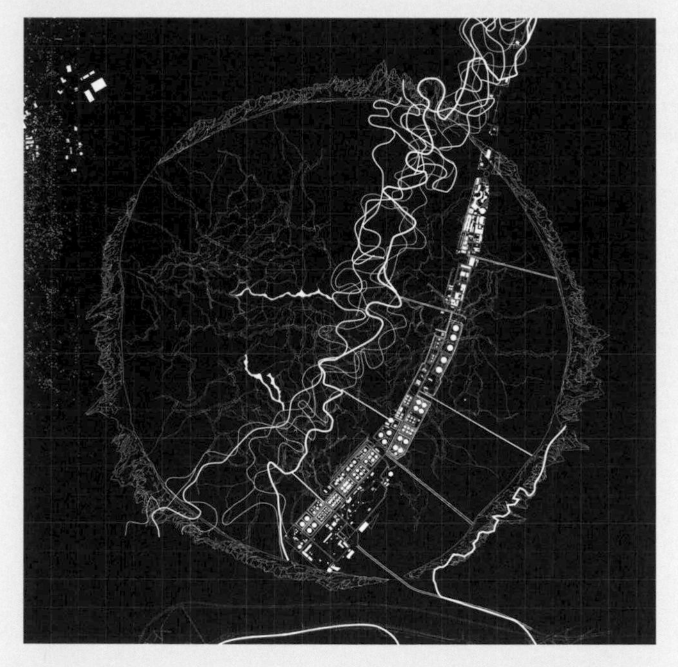

Suction cup 2
A catchment area in an industrialized valley. It could be Chemical Valley or the petrochemical Mississippi River: industries are located along rivers so we can utilize their water, for transport as well as input and outflow. But the current trend is to abandon river transport and renaturalize river. The beds of major rivers are being restored; in any case, it would be impossible to totally constrain them. Climate change is causing an increase in exceptional phenomena (floods, droughts). This territory therefore faces a double risk: risk of flooding and technological risk.

fig. 34 "Industrial biomes": suction cups for extracting energy

Depletion of the territory's resources generates "holes" after activity ceases

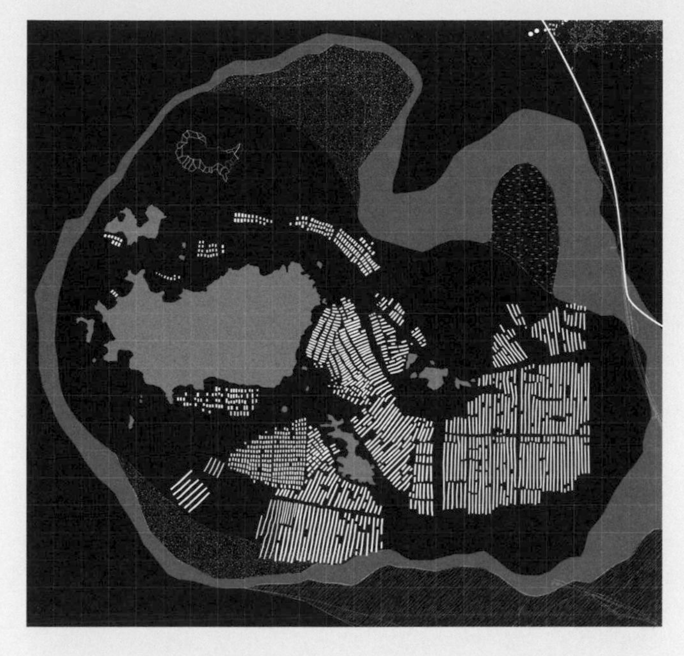

Suction cup 3
A coastal region where fish farming or seaweed cultivation is practiced. Fields in the water range from a few hectares to several square kilometers. Coastal regions are fragile ecosystems because they mark the transitions between land and sea; algae can sometimes be observed suffocating the coasts. But these regions are also threatened by rising sea levels, which could destroy the livelihoods of millions of people.

Suction cup 4
Productive plains. In this space laid out as a large logistics area, resource delivery and storage infrastructures (highway connections, ports) mark the poles while a secondary framework divides the area up into production subunits. Large amounts of water and energy are used to maintain production in this territory. This production area is shaped by the arrangement and repetition of units used for intensive agriculture (greenhouses).

inability to inhabit Earth with other living things. Can we think of any model for our relationship with resources other than that of extraction and depletion? How can we bring nonhumans into the equation, not as resources to be exploited, but as beings with legitimate needs? How can we move from an extractivist and productivist vision of living things and the Earth to a combined strategy between nonhumans and humans? The model that follows does not claim to answer these questions, but it attempts a first visual representation of livable futures.

The future of our resources may be found *between* biomes; we will test this hypothesis here. These areas that extend between the suction cups are never considered as resources and could be a powerful driver of transformation. The map that we have developed here is less descriptive than prospective. It is focused on the enhancement and development of this interbiome area, a living area that includes micro-resources hidden in the interstices of everyday life. By reconfiguring this area, we can experiment with intervening in the dynamics of living processes. This neglected "neutral" area could become an intense area of communication akin to a brain's synapses, destined to replace biomes isolated in bell jars and to develop forms of interspecies collaboration: extensive diplomacy, shared arts of living. A first step in making the drawing is establishing a grid of points dispersed over the whole of the territory, "dynamic sources" of resources that can be activated around, on the fringes of, or inside the depleted sites (see model pages). They bring together artisanal and technical know-how, intellectual inventions, micro-actions, associations, as well as pollinating bees, plants, agricultural plots, and so on: actors who need mediation to become the territory's fabric of wellsprings, its canopy of sustenance. These sources of sustenance need space in which to unfold; depleted territories, for their part, need new springs to reconfigure themselves and connect with life in the rest of the area. The challenge, then, is to connect depleted landscapes and ruins to dynamic sources and living things. The strength of the interbiome, made up of dead points and future resources, lies less in these points than in the links between them. New resources will forge links between things—bodies of knowledge, brains, machines, plants, animals, and so on—constituting hybrid ecosystems. At this stage, we must redefine the notion of habitability. It no longer designates the capacity of a place or a territory to harbor living things, but rather the potential of the entities already present there to make a world together through their interrelationships: exchanges, cooperation, hybridization, additions . . . The space represented on our maps would be at the stage at which these alliances play out; it is a place of encounter.

The second aspect developed on the map is the networking of sources of sustenance through processes borrowed from weaving techniques (see map pages). Unlike the suction cup model, which mangles the warp and weft, this is about reinforcing a fabric that allows for the existence of points and reveals the sources' varying sizes. Several weaving techniques can be used to reestablish continuity: mesh, which creates a new surface; sutures, which allow us to bring two distant surfaces together; the cross-stitch, used to generate a surface in a damaged or perforated place, by increasing its density. With these practical metaphors (weaving, mending, spinning), we seek to summon another register of actions to care for sites that have been rendered uninhabitable. As a method for reactivating points that have been declared "dead," for example, weaving contrasts with applying a patch (as in city planning whose aim is to salvage and repair), which sticks a bandage on a wound or replaces a failed part with a different one. The technique of patching consists of adding an external auxiliary where it would otherwise be necessary to envision a new combination, since in reality holes in a territory are ecosystems that have become obsolete, failing, disorganized, or which have exploded. The portion of the territory that goes unconsidered, the gray area around the suction cups, becomes gray matter, connected synapses, sensitive tissue whose weakened links develop strength to establish a new relational structure. Tying the knots of partnerships, embroidering exchange interfaces, weaving bridges, stitching exchanges—the living tissue will help to extend all these incipient actions.

Tracing this prospective cartography of sources requires detailed knowledge of the territory and the diffuse know-how that covers it, if we are to contemplate its reweaving. Here we are only sketching a map based on a territory in the south of the Île-de-France, between the 13th arrondissement of Paris and the forest of Fontainebleau. The map's design is less a territorial application than a cartography of potential weavings, the various processes, and their ways of forming links. We imagine that this model could allow for the development of open-source maps, where everyone could learn or provide information about a site or project—a kind of know-how or knowledge, with each person acting as a "source."

MODEL VI
(RE)SOURCES

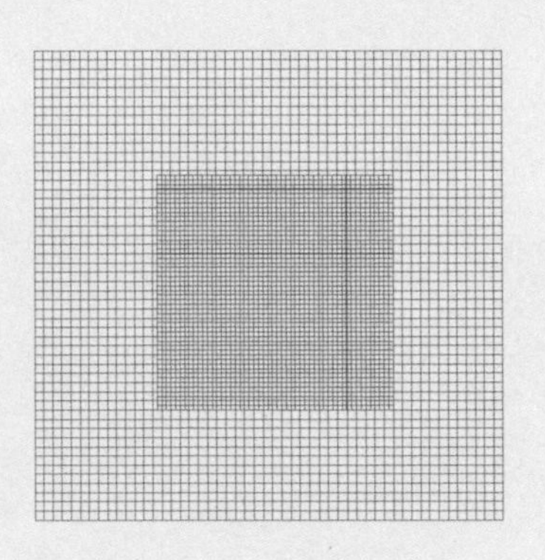

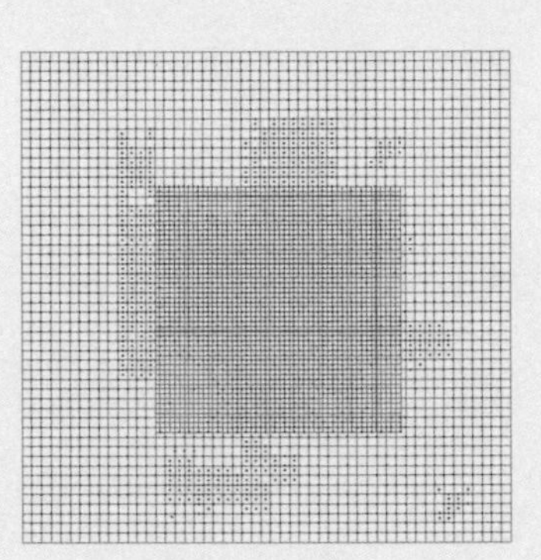

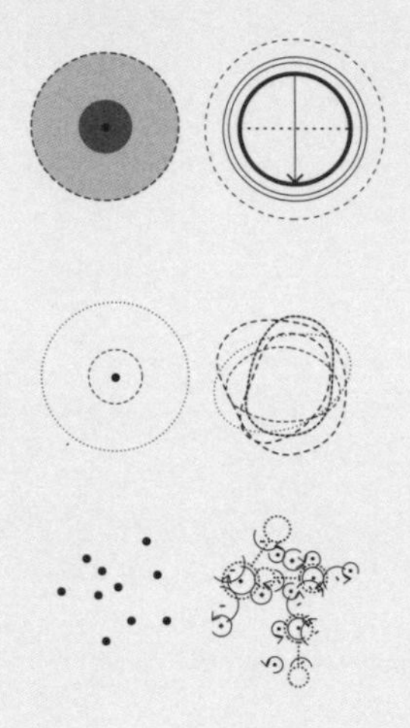

1

Draw a mesh over the territory according to the type of sources sought. A fine mesh will allow micro-sources to appear. A looser mesh is better for locating sources that cover a large portion of the territory. Different layers of mesh can be superimposed.

2

Mark each known or suspected source, whether it is active or inactive, depleted or intact, using a point on the mesh. The sources in a territory are the entities that maintain life in that territory. Some of them are known and easily identifiable; others are dead and will therefore require reconversion; still others are hidden or dormant. This last type raises ethical questions: should they be revealed, how, and for whom?

3

Characterize the source using the drawing rules proposed for each of the three situations: active, depleted, or potential. Active and depleted sources are often linked to the operations of suction cups. The suction cups drain the territory's resources, creating an extracted area. They also generate a halo around this area, a buffer zone between the resource and the rest of the territory. The function of this "phantom" space is subordinate to that of the suction cup. It is phantom because it is dedicated to the functioning of the suction cup but is often overlooked on maps.

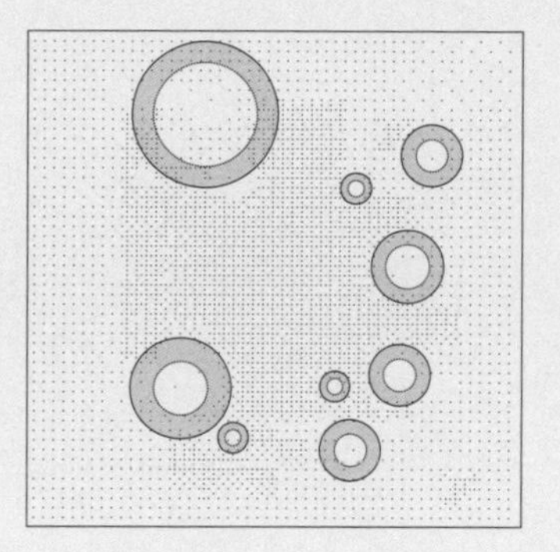

4
Characterize the operation of each of the suction cups into one of the four dominant states: extraction, recovery, isolation, and regeneration.

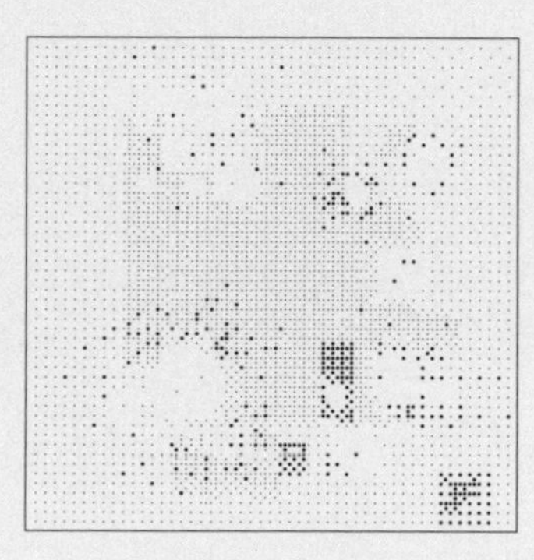

5
Illuminate potential sources near vampirized/encapsulated sources. These micro-sources are the territory's gray matter. The objective, however, is to connect, not to deplete, these sources. The synapse serves as our model: energy is generated through touch and through the connections between elements.

#
+
x
()
:
--
°
',

6
Use the table of operations as a system for combining potentials. In the table of operations, each sign corresponds to a weaving method, a gesture used in sewing.
This translation makes it possible to pair up the operations that generate mesh with the previously identified potential sources.
The variable equations allow us to test the feasibility of the potentials simply and manually before activating them in the territory.
Thus, it is a test program.

MODEL VI
(RE)SOURCES

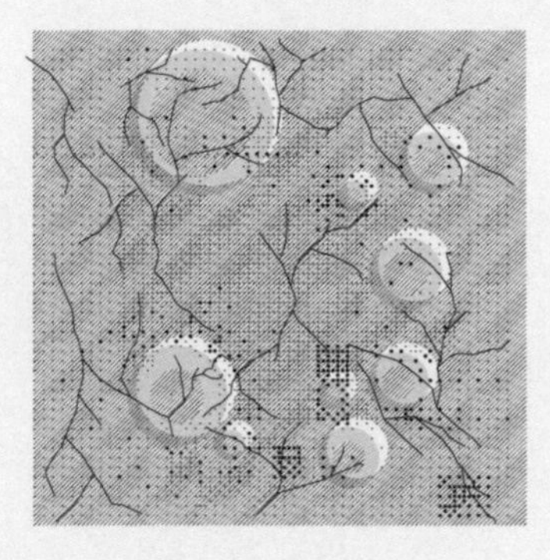

7

Connect the potential sources with each other, entering the suction cup areas from the outside so as to change their state. Thus, the method is to regenerate from the edges, starting from what already exists, rather than from within the closed and depleted spaces.

Suction Cup

Depletion of a resource

Phantom space

Dynamic sources

Adapting the grid according to the territory

Intensity gradient

External sources

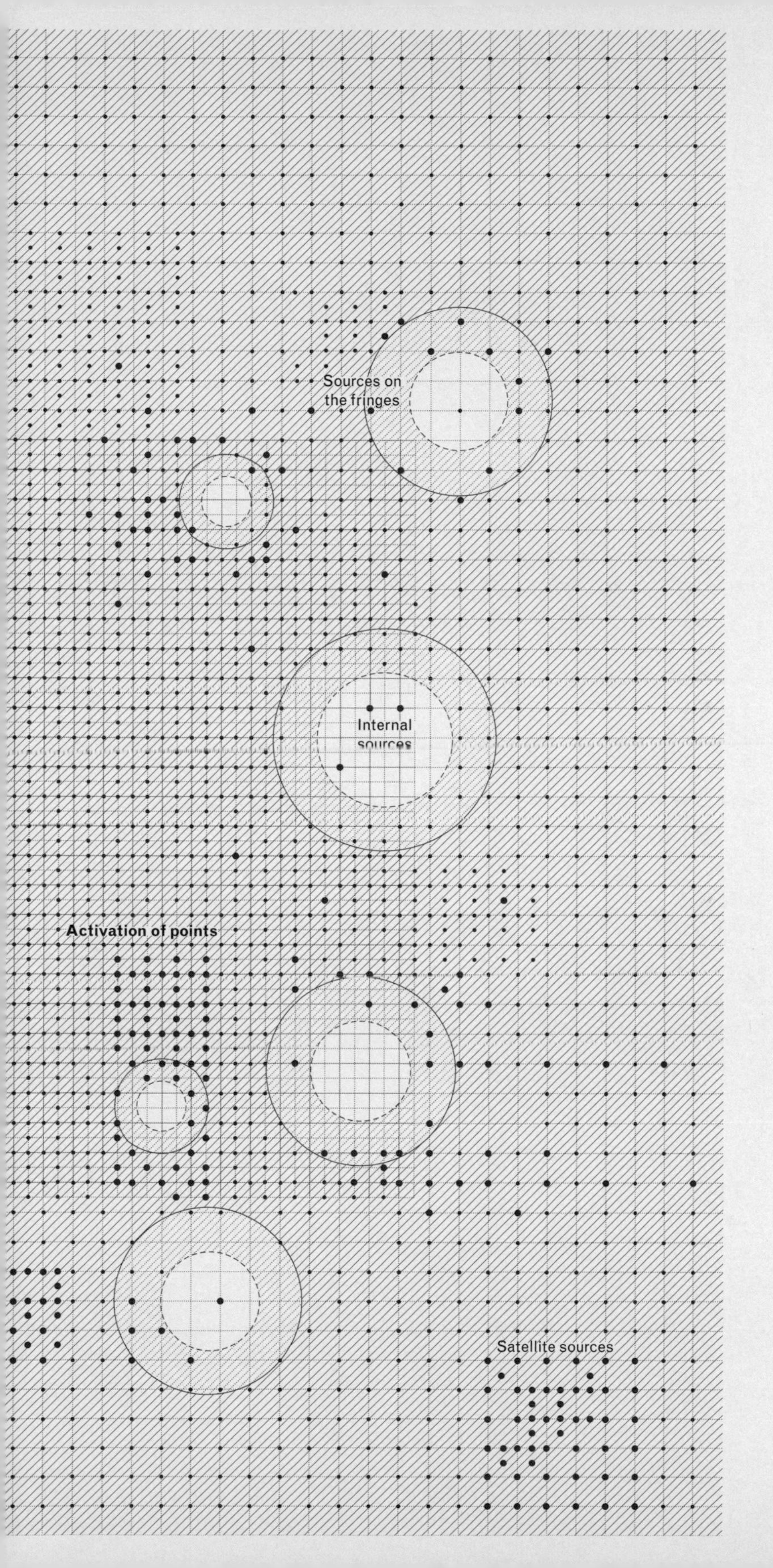

Sources on the fringes
Internal sources
Activation of points
Satellite sources

#

+

x

( )

:

--

°

‘,

MAP VI
(RE)SOURCES

The open-sources map is a canvas to be woven between the suction cups. Weaving allows us to connect different sources that were previously separate: sources such as know-how, open land, commercial or service initiatives, financing, and so on. The modes of association between these sources are imagined as formulas, and are then translated into types of weaving (legend on the opposite page). This operation allows us to create new resources.

For example:

Gardening club × seeds # wasteland (real estate program) × gardeners (insects) + town hall + place of sale × consumers = local food resource
Carpenter craftsperson -- sawmill: architectural agency + abandoned industrial warehouses : real estate transaction = establishment of a timber industry
Former airport site # booth spaces × merchants = market

Where the signs refer to types of weaving:

- # mesh: creates a new surface
- + suture: draws together two distant surfaces
- × cross-stitch: generates a surface in a damaged or perforated place, increasing its density
- ( ) invisible or slipped stitches: generate an underground network and connect points from two different territories
- : back stitches: develop a line on the surface (backward) and below the surface (forward)
- -- continuous stitch: weave connections along a scar

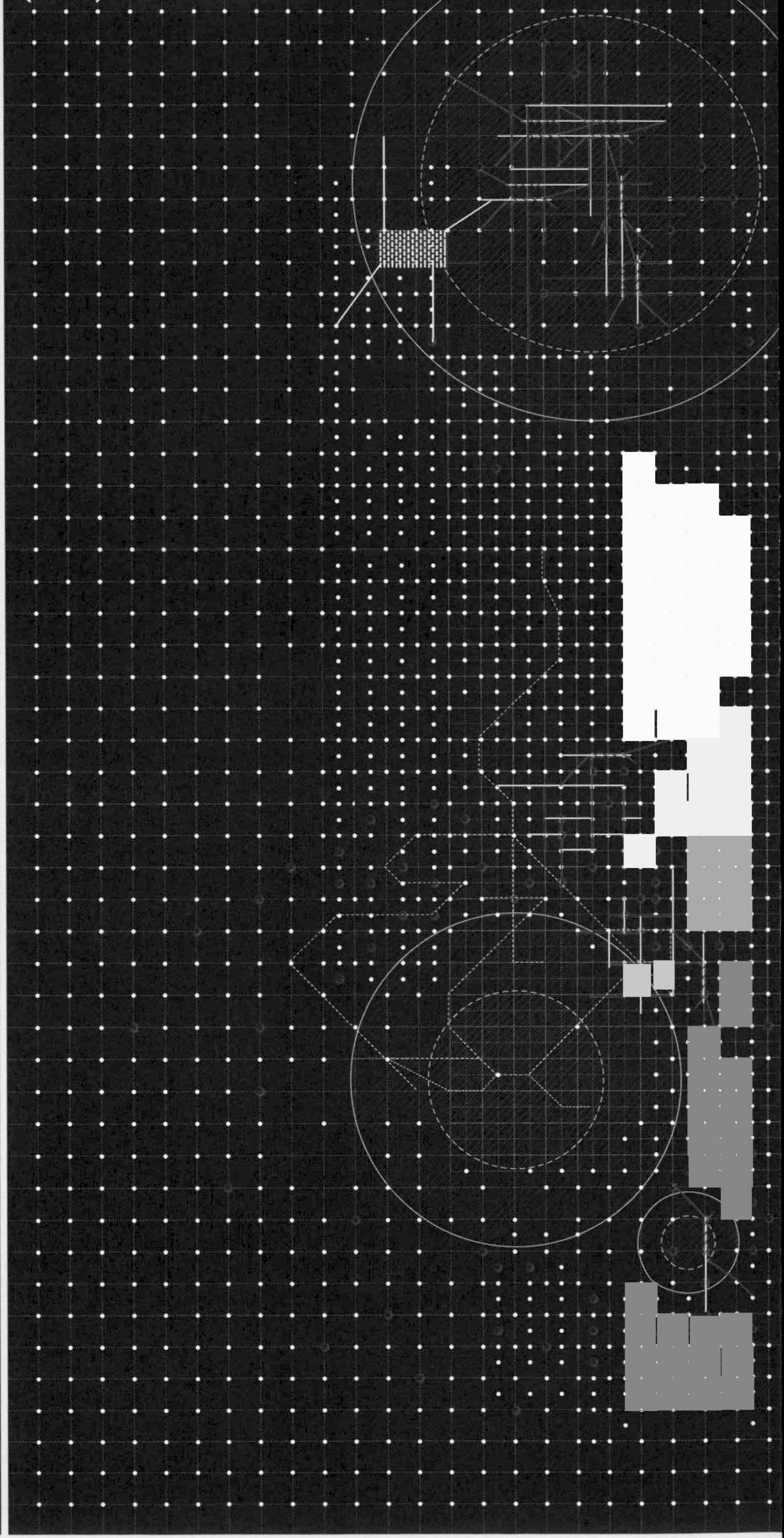

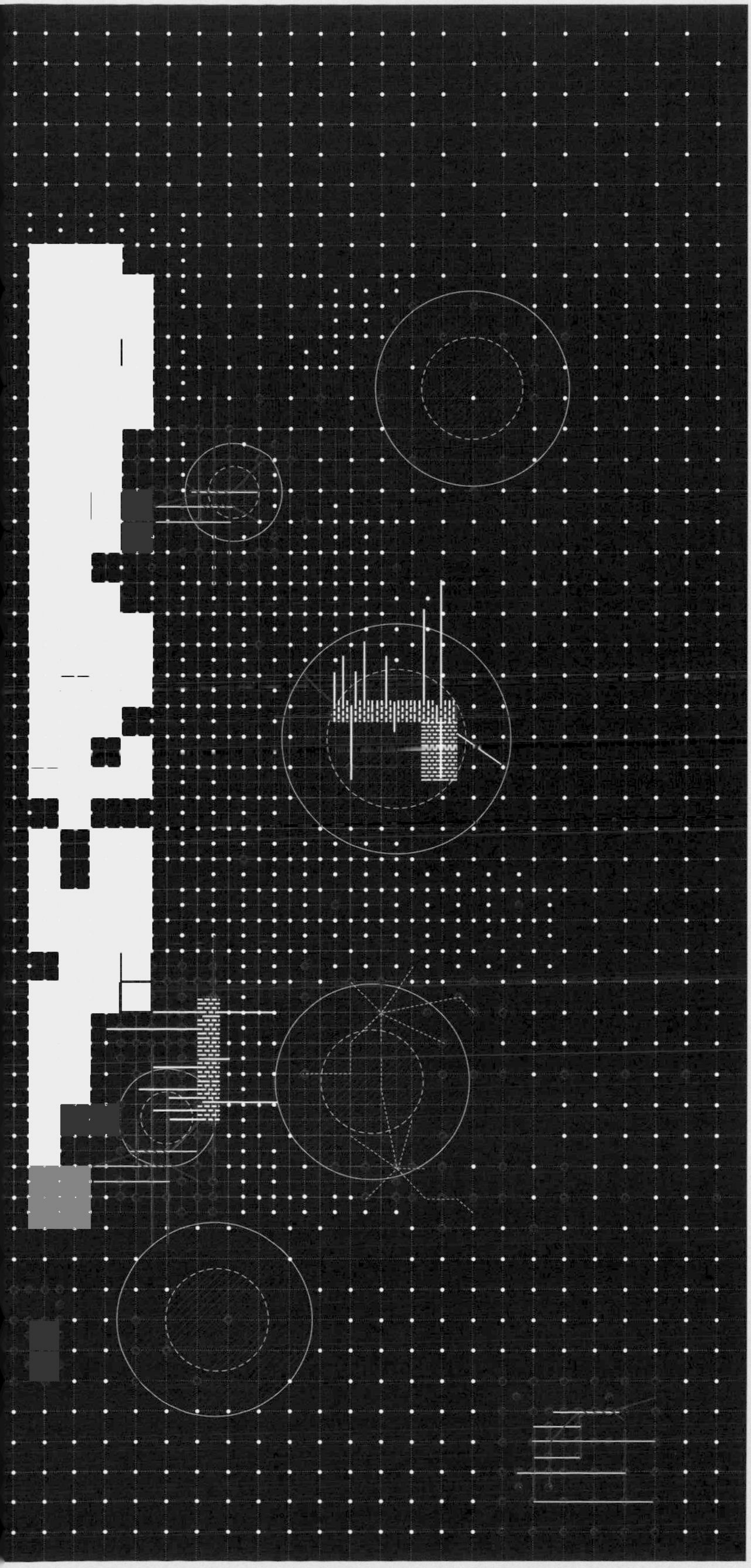

Process of reactivation

Weaving #

Draws together two distant surfaces. From the exterior, the sources' territories, toward the depleted territory

Braid --

Diffusion and thickening
From the heart of the depleted territory

Suture +

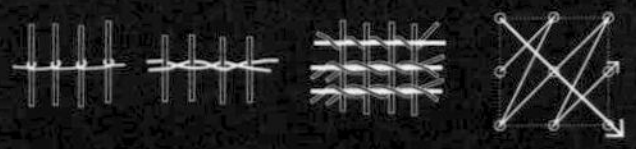

Generates a line or network that connects points from two different "embankments"

Embroidery ×

Generates a surface in one place, increasing density

Knot ()

Creates a new point

TRANSFORMATION VI (RE)SOURCES

From suction cups . . .

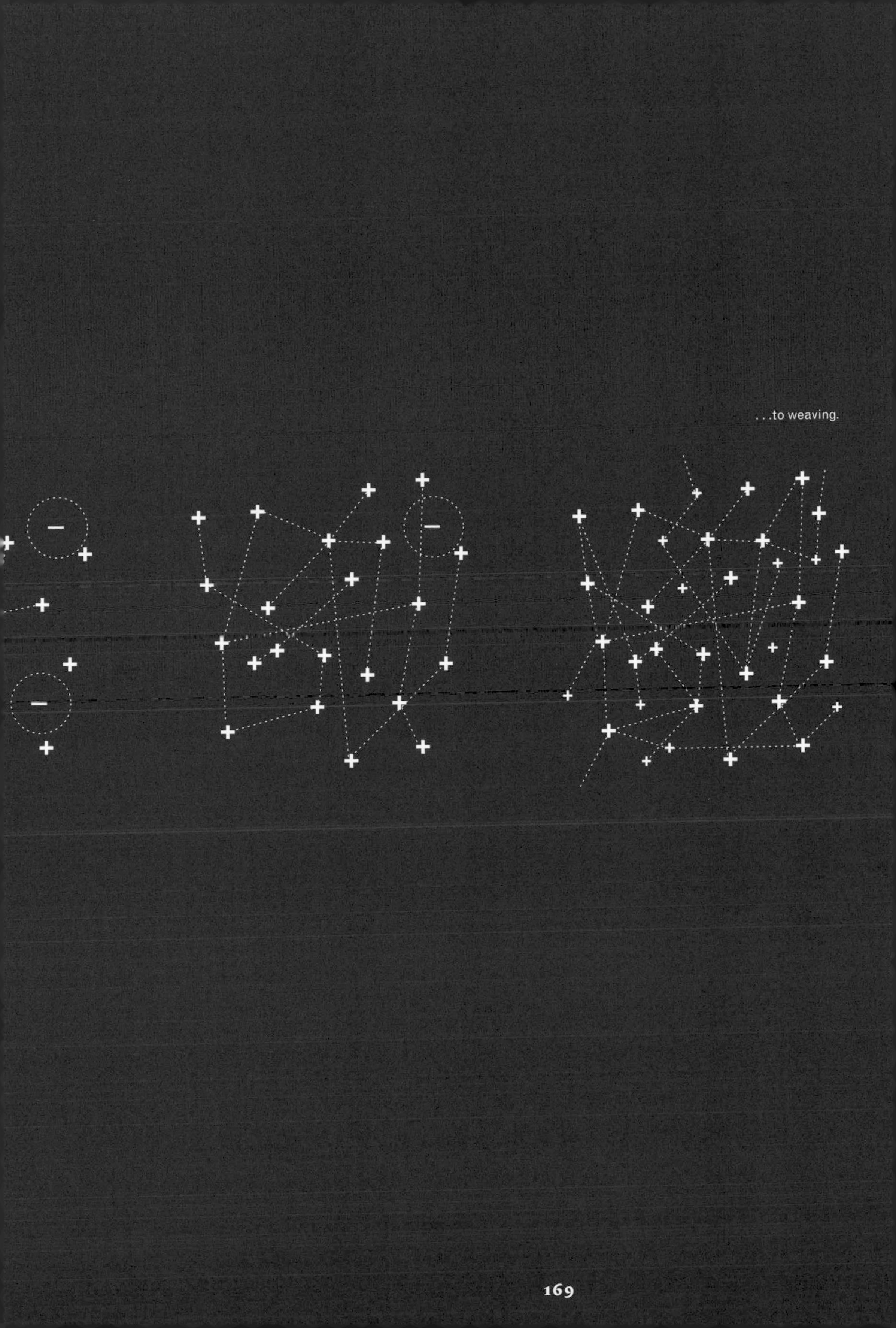
. . .to weaving.

13. Chemical Valley, Grand Lyon, France. Tanks. **14.** The Chapelle Charbon site, northeast Paris, France. Iron landscape. **15.** Chemical Valley, Grand Lyon, France. Phantom landscape.

10. Chemical Valley, Grand Lyon, France. The refinery and steepheads. **11.** Chemical Valley, Grand Lyon, France. Canal as machine. **12.** Hiroshima, Japan. City and memory.

7. Geothermal power station, Iceland. Energy landscape. **8.** Blandan Park before renovations, Lyon, France. Concrete slabs. **9.** Tempelhof, Germany. Former airport site.

4. Chemical Valley, Grand Lyon, France. Industrial landscape. **5.** Odeillo solar furnace, Pyrénées Orientales, France. Energy landscape. **6.** Island of Miyajima, Japan. Old-growth forest.

1. Nara, Japan. Cohabitation. **2.** Blandan Park before renovations, Lyon, France. Plants reclaim the territory. **3.** Chemical Valley, Grand Lyon, France. Infertile and polluted ground.

MODEL VII

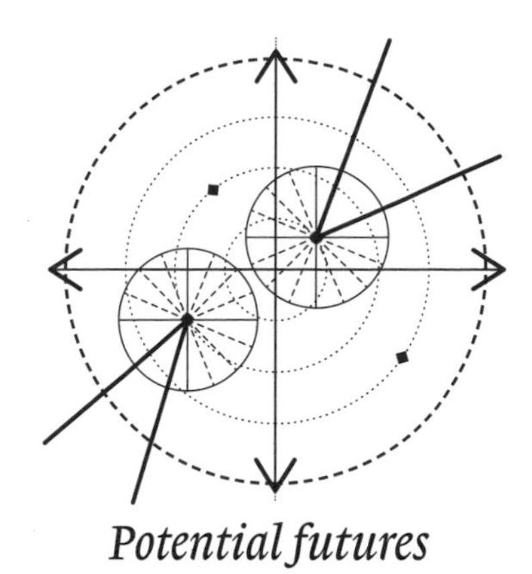

Potential futures

(RE)COLLEC-TION

28. Montpellier-Barcelona, France-Spain. Transborder landscape, forest as machine. **29.** Kobe, Japan. City rebuilt after earthquake. **30.** Chemical Valley, Grand Lyon, France. The chlorine ball.

25. Northeastern Paris, France. Infrathin landscape. **26.** The Pyrenees, France-Spain. Border landscape. **27.** Jökulsárlón, Iceland. Iceberg, landscape in resistance.

22. Chemical Valley, Grand Lyon, France. Pipelines, aerial landscape. **23.** Osaka, Japan. Infrastructure landscape. **24.** Chemical Valley, Grand Lyon, France. Underground pipeline, invisible landscape.

19. Landmannalaugar, Iceland. Ground in motion. **20.** Strengbach Critical Zone Observatory, Aubure, Haut-Rhin, France. Acid rain. **21.** Víti Crater, Krafla, Iceland. Ground as repository.

16. Former coal mine, Søby, Denmark. Industrial relics. **17.** Chemical Valley, France. Former railway site. **18.** Geysir, Golden Circle, Iceland. Active ground.

(RE)COLLECTION

Our journey is coming to an end. We have reached the point from which we started, the ground, which of course we never left. But it is a very different ground from the flat surface generated by the zenith point of view. What we have discovered is thick, granular, reactive, mixed, filled with remainders inherited from the movements and activities of the living things who traverse it, literally, in all directions. In some places it is worn; in some places there are holes. There are vestiges that have become unusable, some of them bulky architectural and industrial fossils—ruins, the subject of our last map. Ghost towns and the too-rapid obsolescence of our infrastructures raise questions about our relationship to destruction, to the entropy that our building systems will necessarily undergo.[44] They also pose questions about the populations who live and will live there. Is the ruin a loss that we have not been able to control, or is it a natural process of reintegration into a cycle, a system of order? Traditionally, ruins indicate a narrative that is coming to an end, an accumulation of traces that no longer make sense. But we could just as well think of ruins as part of a motion that begins with construction and moves to the generation of ground. It is this motion that interests us here: considering ruins as a new topography, assimilating them into natural ground in order to renew their habitable forms.

The phenomenon of ruins isn't limited to well-studied dead cities, commercial and industrial wastelands governed by the obsolescence of our urban infrastructure. There are also invisible ruins: inert, unfertile lands, concealed nuclear and industrial waste, or sand extraction from the seabed, which

---

44 Mathias Rollot, *L'Obsolescence: Ouvrir l'impossible* (Geneva: MétisPresses, 2016).

threatens the equilibrium of the liquid world. The new frame of reference provided by the Anthropocene is a revelation that makes visible the invisible. It gives rise to ghosts, as well as emerging lobbies. One example is the mushroom "at the end of the world" described by Anna Tsing, which appears as a forerunner on sites that have experienced disaster, the mycorrhizae transforming the chemistry of the soil to make it fertile and habitable again. The idea is no longer to highlight the ruins as Robert Smithson was able to do, but to make them visible through the development of another cartography, an invitation to explore a new *terra incognita*.

Like an organism, a territory has DNA, a memory that encodes a mass of identifying information, traces, and imprints. This mass can be inert, buried under piles of ruins, or it can be activated. The degree of a territory's habitability is often measured by its inhabitants' ability to bring this memory to life—memory that, like ours, is largely fabricated, cobbled together, hybridized. Some traces are accentuated, others are covered up. The remains, fossils, and relics of the past are sometimes staged, as in a memorial. More often, we try to make them disappear. Lower layers of soil occupy the paradoxical place of both a material and collective unconscious used to bury what we no longer want to see. We thus readily forget the past of certain sites without assessing the problem of their traces: depleted ground, for example, is dead ground, an invisible but cumbersome ruin. Nuclear waste and infrastructure networks are buried underground so as not to impede the movement of living beings but also, in part, so that we can forget them. It must be possible to map the ruins of geohistory, this very young human geology made from the gradual sedimentation of our artifacts.

We hope, following in the footsteps of Anna Tsing, that these traces can be the foundations of a possible future. In this chapter, we have imagined a more prospective tool that uses waste from the past as a material to be recycled for manufacturing futures, allowing transformations. Designed as a device for connecting a past and a future, the model poses questions about the processes of activation, transfer, encoding, or inversion that could make it possible to live in these ruins (see model pages). It is less about reading the past than thinking about new foundations, less about the archaeology of disaster than about living memory. Human traces mark the territory, but reading it is difficult. For this palimpsest soil to beckon to us and allow a form of reading and interpretation of territories, we must be able to encode what we want to preserve in memory: to build a semiotics of ruins. Our model provides a rough outline for this proposed production site of signs. We will start by collecting and archiving the most prominent fossils, the most visible ones, and those that pose a problem despite or because of their invisibility. We will collect them as a means of reclassifying the landscape differently and of thinking differently about these artifacts that are rarely considered together. Archiving these traces allows them to turn them into semiophoric objects, endowed with value and meaning, and to begin to constitute a library of ruins, materials and forms that we can draw upon to construct future scenarios. The production site of signs thus invents a new grammar: reinterpretation, erasure, transfer, inversion. It lists the processes (creation of sanctuaries, renovation, recycling, reinterpretation, transformation), considers the potential uses of a form (habitation, construction), and suggests rearrangements. In short, it imagines habitable ruins (fig. 35).

But this desire to occupy ruins is in conflict with some of them. Today, we are increasingly confronted with sites of heavy industry where the morphology of the territory has totally changed, and geological history merges with industrial history. What should we do with this heavy heritage, with toxic landscapes and dangerous ruins? Some traces correspond to breaking

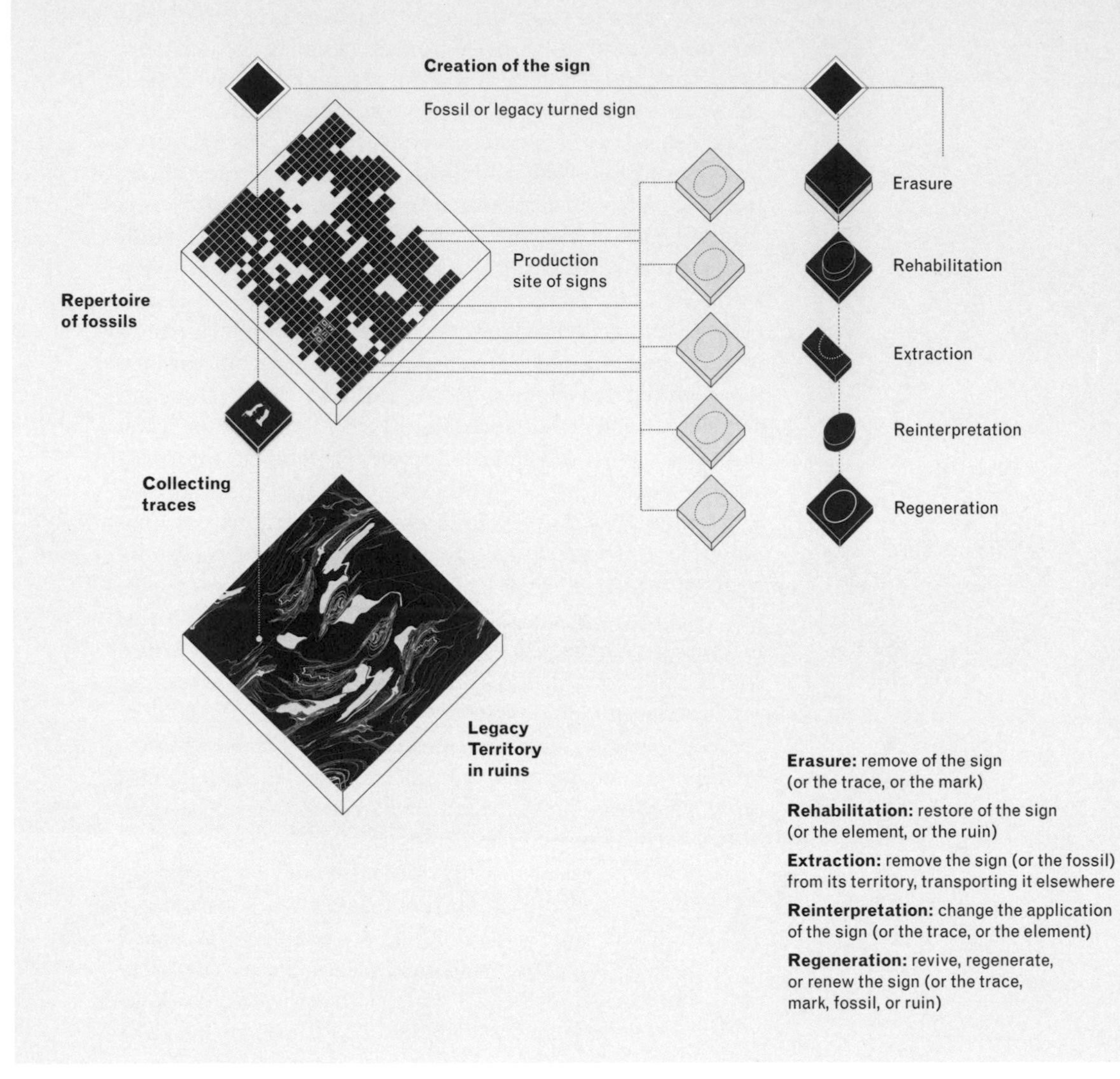

fig. 35 The production site of signs

points, at the radical end of the habitability of an area. These impose a dead time of variable duration, exceeding, in the case of nuclear waste, the temporal scale of human history. But other than this extreme case, most ruins can be rehabilitated. Each case opens up a possibility. For example, architects are fascinated by bunkers. Some have gone so far as to reinvent them as architectural "follies," in the process helping to reevaluate concrete as a building material. The saw-toothed roofs of industrial buildings have become a marker of arts organizations almost everywhere, in a successful reinterpretation of signs. If the ruin as cultural heritage is less interesting (it is simply put under a bell jar), recycling and reinterpretation, in contrast, go hand in hand with a wide variety of potential scenarios. The map thus becomes a political tool for envisioning the future, a kind of compass for orienting oneself (see map pages). It takes the form of a toposcope equipped with a library of ruins, a sign encoder, and a simulator, opening up clusters of scenarios inside the map.

The cartography in these pages is multisite. It is inspired by the territories of migratory animals and temporary camps built, destroyed, and dreamed of; nuclear waste burial and treatment sites; soil marked by war, combat, shells, bombs, and buried bodies; or, less dramatically, it cites industrial and agricultural wastelands, ruined territories, and environments deprived of their ecological functions. It is necessary to imagine a possible life on top of these ruins, to project a habitable future, not in the classic cartographic sense of a projection in space, but in the sense of a projection in time, in order to patiently sketch a political compass made from the debris of an astounding collection of fossils.

MODEL VII
(RE)COLLECTION

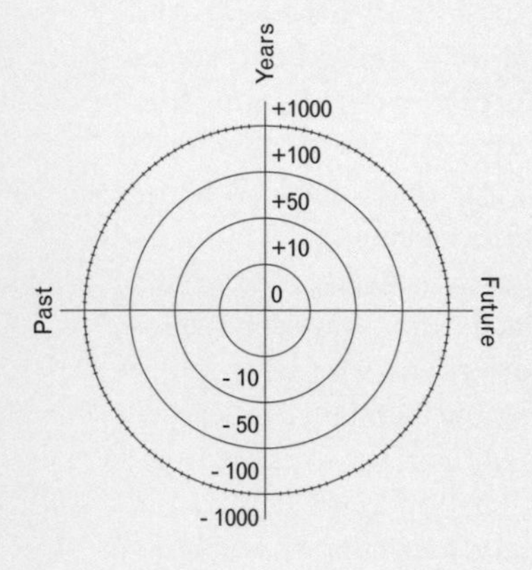

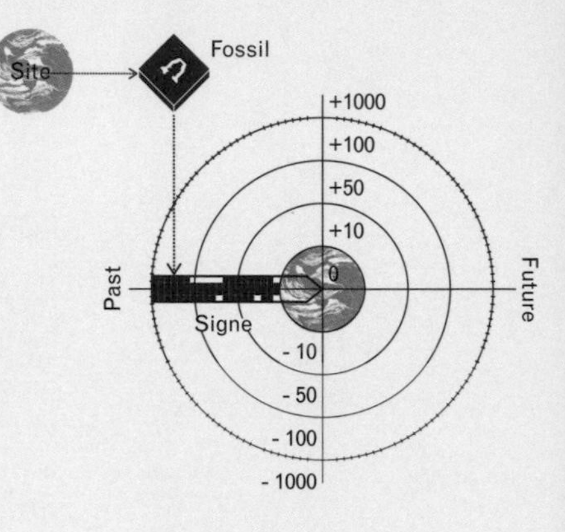

1

Generate the compass. Inset circles form a target. Each circle corresponds to a date in the past or the future calculated from date 0 inscribed in the center. We thus obtain a sort of temporal thermometer. The right half of the target shows future times (+5, +10, +50, +100, +500, +1,000 years), while the left half of the compass shows past times (–5, –10, –50, –100, –500, –1,000 years). Date 0 in the center is the date of manufacture of the map based on this model.

2

Collect traces from the territory. These traces, ruins, or foundations are called "fossils." To establish a repertory of fossils, extract the pertinent elements from the site, future "signs," and turn them into samples. This repertory will be enlarged from sources other than the studied territory and can thus be consulted and activated for other territories.

3

Place the studied territory at the center of the compass by depicting its ground's geological layers. The layers might be natural or artificial, ranging from rock to constructions. Classify the fossils or remains of this territory in the historical timeline based on the compass's time gradient, assigning an age to each fossil.

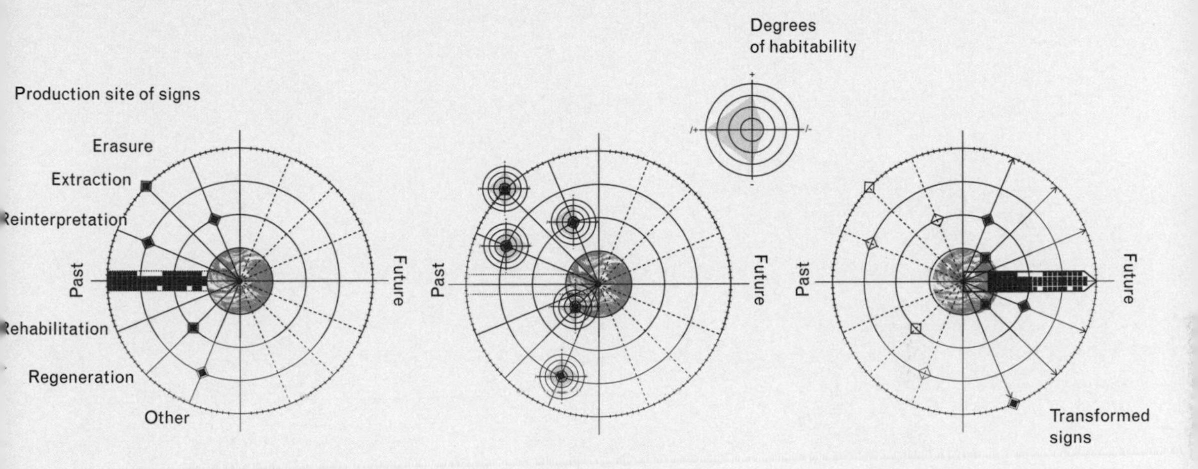

4

Starting from date 0/the current state of the site, draw six rays on either side of the axes of past and future. On the six rays in the past portion of the compass, write down the operations of transformation expected to change the sign into potential for reconversion. These actions correspond to what is called the "production site of signs." Our model contains six of these actions: erasure, rehabilitation, extraction, reinterpretation, regeneration, and other (for special cases). Place each of the territory's signs at the intersection of a ray (an action that causes transformation) and a temporal arc (the age of the sign).

5

Study the territory's degree of habitability based on each sign: does the element have the potential for rapid or slow reconversion? Studying the habitability of the ruin will allow us to determine whether the transformations are feasible and what their timeline for realization might be.

6

Move the transformed signs to the part of the timeline marked with arrows on the right side of the compass: this will be the future timeline. Position each of the signs along the compass rays that are a continuation of their past, according to the planned action and the expected timeline for the transformation of the site. We thus obtain a mirrored structure from which we can create a visual representation of the number of years necessary for the transformation of the fossil into a sign, allowing the site to be habitable again.

MODEL VII (RE)COLLECTION

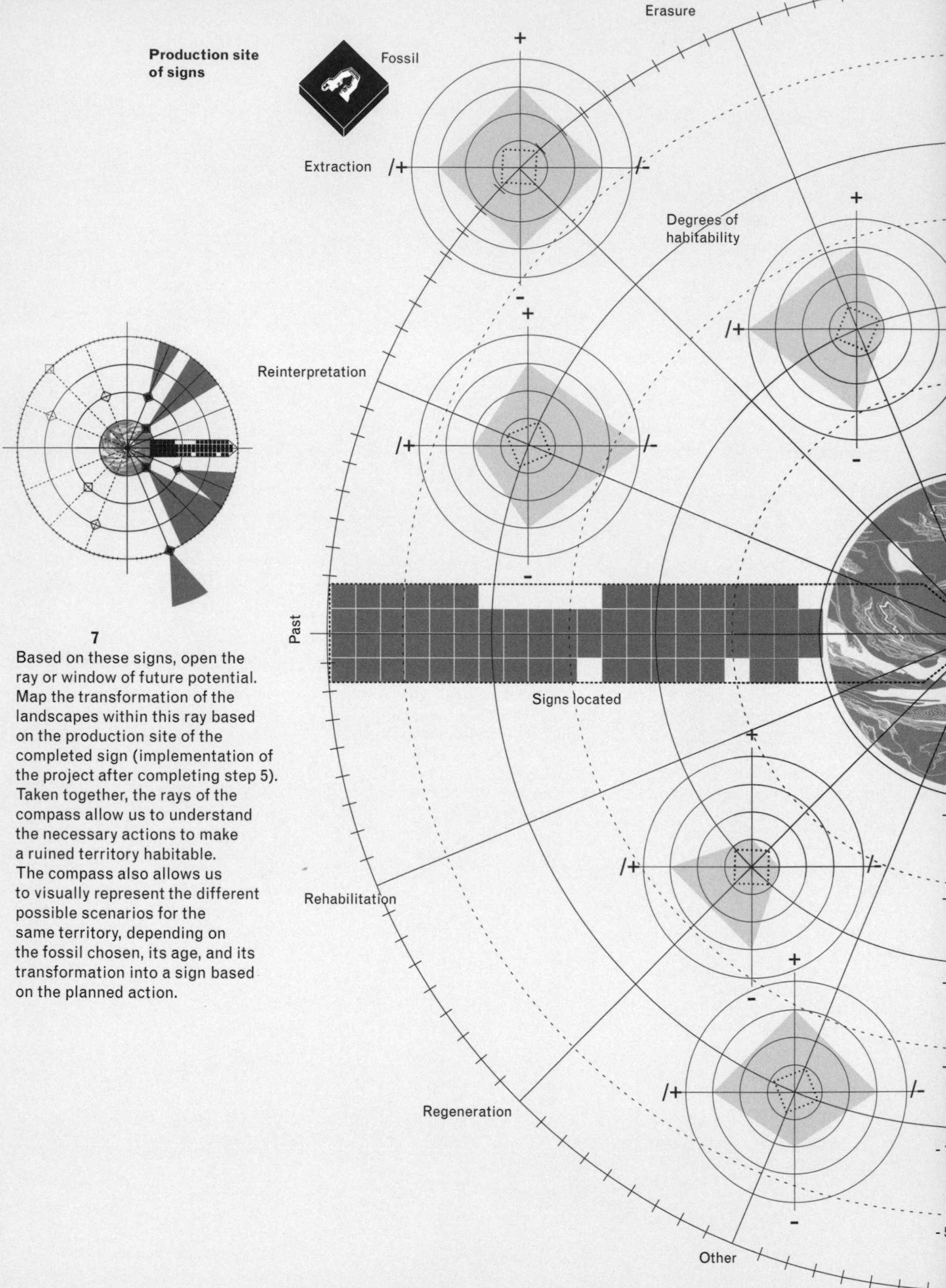

7

Based on these signs, open the ray or window of future potential. Map the transformation of the landscapes within this ray based on the production site of the completed sign (implementation of the project after completing step 5). Taken together, the rays of the compass allow us to understand the necessary actions to make a ruined territory habitable. The compass also allows us to visually represent the different possible scenarios for the same territory, depending on the fossil chosen, its age, and its transformation into a sign based on the planned action.

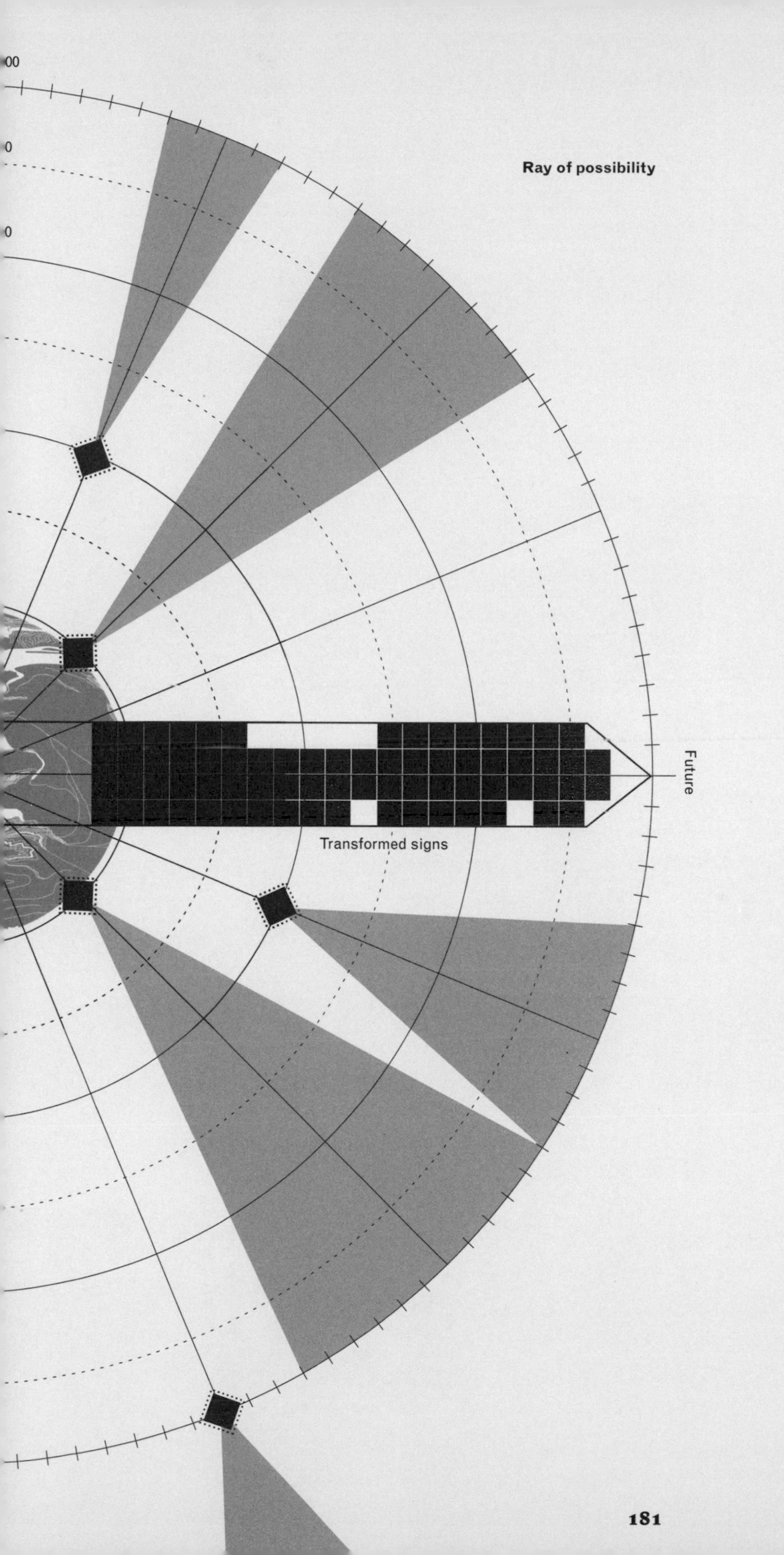

00
0
0
Ray of possibility
Future
Transformed signs

The compass of possible futures based on ruins
Typology of fossils and their age

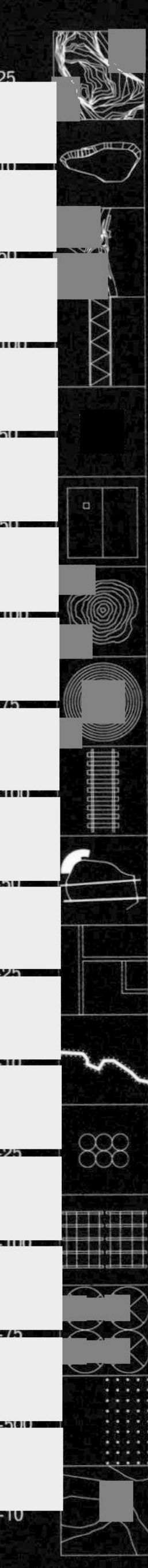

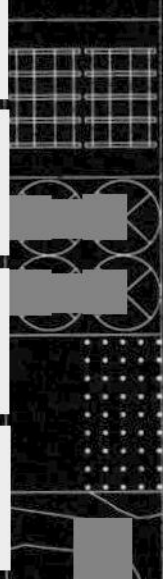

MAP VII (RE)COLLECTION

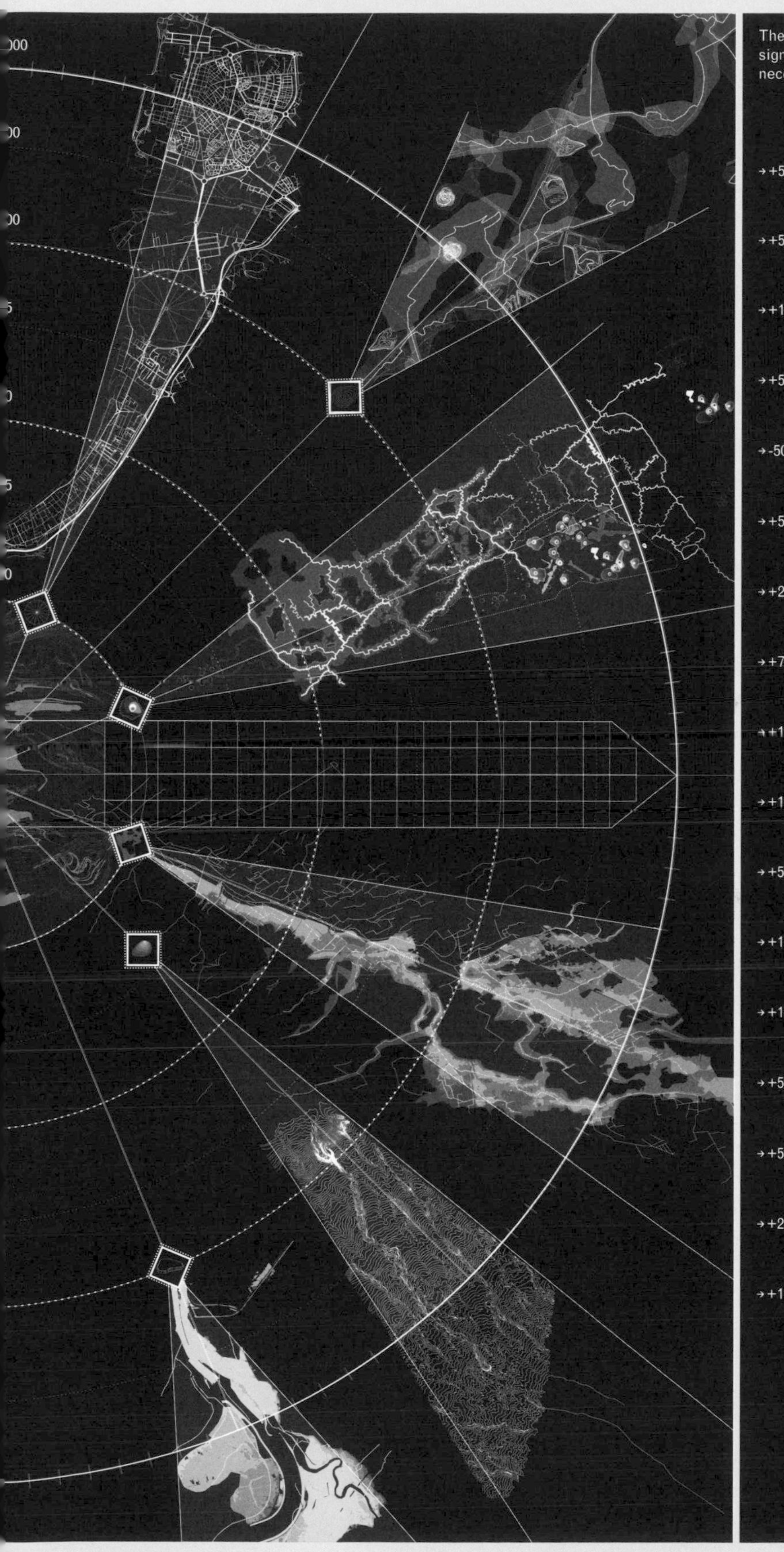

The same fossils transformed into signs, with an indication of the timeline necessary for their transformation

Timeline	Sign
→ +50	Mountain
→ +5	Water reservoir
→ +10	Wildlife corridor
→ +50	Islands
→ -50	Factory of fertile soil
→ +5	Animal city
→ +25	Anthropocene park
→ +75	Museum of physics
→ +10	Ghost farm network
→ +10	Urban refuges
→ +50	Cooling well
→ +100	Transborder forest
→ +1000	Memories of the nuclear
→ +50	Rotating market hall
→ +500	City on stilts
→ +25	Stand of trees in a farmfield
→ +10	Fertile spots

DRAWING AN UNKNOWN WORLD

We unfold a map. Our eye recognizes lines, orients itself, and constructs the space. To moor ourselves, one finger rests on the paper, transporting the body, causing it to travel into the authorized space of wandering: public space, the landscape. But the body soon collides with the boundaries of parcels and blocks of construction, restricting it to what is walkable, taking roads, not daring to venture into the blank space that represents an entirely different mosaic. Our visual representations of the Earth are based on ancient cartographic and geological images. Designed as "a tool of colonization, a way of writing the story of a conquest where the civilized takes possession of so-called 'empty' territories, but that in fact it is always a process of 'emptying,' because they are populated."[45] Populated by organisms, commemorative traces, industrial elements, rocks, bacteria, tunnels, water, volcanoes—a conglomeration of living organisms and hosted objects buried beneath our feet. In this heterogeneous soil, the administrative boundaries on the surface no longer determine the map; instead, the moving body of living things do. "It's a map that seems to reflect a liquid space . . . where routes are marked until they are blown away by the winds. This nomadic space is crisscrossed by vectors rather than paths, unstable arrows which constitute temporary connections."[46] By imagining a type of documentation in which the paths are not erased but are inscribed and printed, where they accumulate, overlap, and mingle, the models proposed here attempt to capture the contours of a completely different space.

---

45 Jean-Baptiste Vidalou, *Être forêts: Habiter des territoires en lutte* (Paris: Zones/ La Découverte, 2017), 28–29.

46 Francesco Careri, *Walkscapes*, 43.

Entering the Anthropocene forces us to rethink our relationship to the ground, to navigate the "liquid space" described by Francesco Careri—a moving and unpredictable world where islands disappear, species go extinct, ecosystems are desertified, wars break out over water, seafronts are submerged, and valleys are flooded. The ruin of territories brought about by the climate crisis causes a crisis of space that renders obsolete any physical survey based on old parameters and forces us to question our descriptive tools. In fact, a major source of the skepticism or indifference toward the new climatic regime stems from the lack of a common understanding of what it means to live in these critical zones. The lack of a common vision and imagination is a major obstacle to understanding the roles and interactions that emerge in devastated environments. The new frame of reference offered by the Anthropocene reveals the reality of our situation, encouraging us to change our perspective and reread our territories through the prism of *habitability*. It is this movement that our maps hope to support. By working on experimental representations, diagrams, models, outlines, and maps, we can make visible the invisible (anthropogenic destruction, ruined ecosystems) and explore other ways of locating ourselves in the world. The map transforms into a laboratory and takes on the character of prototyping, since it must make it possible for us to test out possibilities of social, geological, and climate configurations.

The proposed cartographic tools emerged from a desire to overcome the obstacles inherent in the classic practice of land-use planning: discrepancies between temporalities; administrative limits that contradict the needs of geographic entities; real estate being valued more than its soil substrate; a lack of openness between disciplines; biases in memory production; an overemphasis on standardization; the inability to consider the micro-scale, whether in reference to biological entities or the use of resources; and a failure to take into account the conditions necessary for maintaining a functional ecosystem. Issues such as these inspired these models, and the underlying desire to encourage collaboration and disciplinary decompartmentalization. Our choice of a 2D approach was a vote for simpler tools and rewriting processes, which use simple geometry, references to traditional explorers' maps, and drawing as a method of surveying territories. At the same time, on our websites,[47] during participatory workshops, and through public performances,[48] we reflected on how the stage and video would add a new parameter of movement and three-dimensionality . These stage experiences and participatory workshops were steps in our journey, opportunities for testing tools that were still in the prototype stage. This leads us to think today about what the development of these models in the form of software could add. *Terra Forma* has indeed made it possible to define the parameters of a "machine" to generate maps. As for how to ultimately turn the legend into computer code for the (Re)Sources model, we are considering it as a way to exponentially expand data collection, to include new parameters through 3D, or to broaden the dissemination of means of action. Thus a new figure of the cartographer in the digital age emerges, the cartographer-programmer, with the hope that these cartographic and digital tools will also be political tools, making it possible to take ownership of the drawing and design of territories by all who shape them—elected and nonelected officials, activists, citizens, architects, landscapers, public and private investors, farmers, forest

47 See http://s-o-c.fr/ and http://zonecritique.org/.

48 *Inside*, a conference-performance with Bruno Latour, directed by Frédérique Aït-Touati, with maps, visuals, and videos by Alexandra Arènes, Axelle Grégoire, and Sonia Levy (premiered at the Nanterre-Amandiers Theater in November 2016; performances at HAU, Berlin, September 2017, at Pershing Square Signature Center, New York, October 2018, and at Kaaitheater, Brussels, November 2018); *Nouveaux Cahiers de Doléances* (Théâtre Nanterre-Amandiers, May 2018); *Back to Earth*, a participative cartographic performance (Paris, Centre Pompidou, January 2019).

wardens, associations, communities, ecologists, soil scientists, botanists, and geophysicists. At the dawn of a global reinvention of our relationships with living things, a double redefinition is thus carried out: that of cartography as a mode of *potential* reading and writing; and that of architecture as an effort to construct tools for connection and sharing, rather than a desire to build the world. The mission of the architect-cartographers then becomes to manufacture tools for describing the terrestrial, which is in the process of being redefined.[49]

The Earth is a skin. We have drawn its points, traces, holes, scars, excavations, cracks, frostbite, and sunburn. The Earth is liquid: volatile, fluctuating, and protean, like the living things that constitute it. Living things leave traces and signs. They construct territory by moving through it and inhabiting it. These traces need to be deciphered, retranslated, and narrated, like the territories they outline. As such, they invite us to redirect our interpretative skills toward the ground, to relearn to see using new images and other artifacts. To see the Earth as a planet to be explored is to reconnect with the wonder of the Age of Discovery, because this world that is endlessly terraformed by living things remains largely unknown. It is a world of amazing beings, surprising assemblages, and new alliances. Unlike our former Earth, made up of territories to be conquered and resources to be possessed, this one has no limits. It invites us to undertake a quest that cannot be exhausted or depleted.

---

49 Along with architect-cartographers, we can mention writer-geographers, philosopher-trailfinders, anthropologist-writers—an entire family of thought associated with the redescription of shared territories.

APPENDIX: DATA

Our data were first collected at an urban planning and landscape agency, at the places where we studied, and in projects we participated in. These data enabled us to construct the conceptual diagrams that we call "models." As we generated these models, local data fell by the wayside to allow us to develop more general tools. Thus, the model can be used on other sites that the reader would like to address. The model's potential is then explored in a visual representation that we call a "map" (on a black background). These "maps" combine data collected from several territories or from just one. They are always generated from existing elements. They differ from a traditional map in that these elements of the territory are mapped by the means described in the model, which allow them to be read differently.

The models were therefore created partly in reaction to shortcomings encountered in land development projects. The landscape architect, however, recognizes that landscapes are dynamic processes in continual transformation. One of the reasons for this conception of landscape is found in the type of the sites where landscape architects and urban designers are asked to intervene. Most development projects consist in rehabilitating former (usually polluted) industrial sites through processes of land restoration; organizing urban development on devastated, postindustrial or post-exploitation sites; or reclaiming natural spaces (buried waterways, former agricultural sites, former industrial sites, former quarries, former military sites, former railway sites, former ports, former airports, etc.)—those that Anna Tsing rightly calls "blasted landscapes."

I. SOIL

Territory: multisite

The Soil map is composed mainly of elements extracted from the territory of the Chemical Valley in Lyon, from typical former industrial sites, and from data collected from the Critical Zone Observatories network (OZCAR). Other secondary data were introduced to the map to demonstrate the drawing potential offered by the model.

II. POINT OF LIFE

Territory: multisite

Data from the Point of Life model come from extremely varied scales, since the aim of this model is precisely to map a territory no longer at a single scale, but rather simultaneously with all the scales that locate a point of life. The visual representations that are assembled to account for the point of life's habitual terrain therefore extend from a drawing of the skin to the representation of meteorological effects, passing through many intermediate dimensions, which are different for each of us.

III. LIVING LANDSCAPES

Territory: southeast Grand Paris—urban and suburban

This territory is located at the junction of the highly urban and the countryside, with density gradients characteristic of the transition between the two. The idea is to offer the greatest possible diversity for studying the variation, or lack thereof, of figures by residential area (whether human or animal). Grand Paris is the area par excellence for major urban projects, thanks in particular to the establishment of the Grand Paris Express. We wanted to highlight a gap in relation to these large-scale publicized infrastructures and the fragility of our figures of movement, hidden and overwhelmed by the multitude.

The data required for plotting the study were collected from a sample of about ten inhabitants and eco-ethological data.

IV. BORDERS
Territory: the Pyrenees, the archetypal border

In the past, Axelle Grégoire has worked on this transborder territory between France and Spain. Her knowledge and the cartographic data available motivated our decision to map this territory with the Borders model. It is also an interesting territory because it literally physically contains different types of borders, and because its history as a border territory has made it rich and diverse.

V. SPACE-TIME
Territory: northeastern Paris

For the Space-Time model, we chose a district that we studied for nearly four years at the Base agency in a group led by the François Leclercq agency: the northeast district of Paris (the 18th, 10th, and 19th arrondissements) between the Gare de l'Est and the Gare du Nord, and reaching to the ring road. This territory forms a triangle whose edges are the Canal de l'Ourcq to the east, Boulevard Barbès to the west, and the ring road to the north, the tip being the "bipolarity" of the stations (a term used in our urban study); the railways are integrated into this territory as a rupture to be erased or transformed. One of the main constraints (and assets) of this district is the layering of different rhythms of life in the same public space: the stations and their tumult, the markets, the residents, the people passing through, and so on. This layering generates conflicts in places of friction where rhythms collide because they are unable to understand each other, agree, or at least make room for each other. We used this situation as a starting point to build the model.

VI. (RE)SOURCES
Territory: multisite

This map is made up of elements from several formerly exploited territories (industrialization, intensive agriculture, forestry, storage sites, and so on).

A map testing the potential of the model could be located on any territory exploited by "suction cups." Creating a cartography based on this model requires detailed knowledge of the territory and the diffuse know-how that blankets it, in order to contemplate its weaving. The map's design is less a territorial application than an inventory of potential weavings, the different processes, and their ways of forming links. In fact, we imagine that this model could allow us to develop open-source maps where everyone could learn or provide information about a site—a project, a kind of know-how, a knowledge, with each person acting as a source.

VII. (RE)COLLECTION
Territory: multisite

This last map was conceived around types of territory rather than from a series of sites. This body of work focuses on a single typology, that of ruined landscapes, since this model involves rethinking methodologies for projects based on what remains (waste, debris, traces, monuments). Along with data reused from the other models, there are additional surveys, drafted during our research or simply from the collection of ruined landscapes (photos of sites, aerial images, land registry maps). This final chapter is thus at once a conclusion and a new beginning.

BIBLIOGRAPHY

Aït-Touati, Frédérique. *Fictions of the Cosmos: Science and Literature in the Seventeenth Century.* Chicago: University of Chicago Press, 2011.

Alpers, Svetlana. *The Art of Describing.* Chicago: University of Chicago Press, 1983.

Arènes, Alexandra. "Tracer les vivants, Cartogenese du territoire de Belval." *Billebaude*, no. 10: "Sur la piste animale," June 7, 2017.

Arènes, Alexandra, Bruno Latour, and Jérôme Gaillardet. "Giving Depth to the Surface—An Exercise in the Gaia-graphy of Critical Zones." *Anthropocene Review* 5, no. 2 (2018): 120–135.

Atlas of places. http://atlasofplaces.com. Accessed March 5, 2021.

Azam, Geneviève. "Réduire le vivant pour le fabriquer?" In *Les Limites du vivant*, ed. Roberto Barbanti and Lorraine Verner, 367. Bellevaux: éditions Dehors, 2016.

Bellanger, Aurélien. *Le Grand Paris.* Paris: Gallimard, 2017.

Bertin, Jacques. *La Graphique et le traitement de l'information.* 2nd ed. Brussels: Zones sensibles, 2017.

Besse, Jean-Marc. *Les Grandeurs de la Terre: Aspects du savoir géographique à la Renaissance.* Paris: ENS Editions, 2003.

Besse, Jean-Marc, and Gilles A. Tiberghien, eds. *Opérations cartographiques.* Arles: Actes Sud / Ecole Nationale Supérieure du Paysage de Versailles: 2017.

Béziat, Julien. *La Carte à l'œuvre: Cartographie, imaginaire, création.* Pessac: Presses universitaires de Bordeaux, 2014.

Bonneuil, Christophe, and Jean-Baptiste Fressoz. *L'Événement anthropocène: La Terre, l'histoire et nous.* New ed. Paris: Points, 2016.

Bosseur, Jean-Yves. *Du son au signe, histoire de la notation musicale.* Paris: Éd. Alternatives, 2005.

Brotton, Jerry. *Une histoire du monde en 12 cartes.* Paris: Flammarion, 2012.

Brunet, Roger. *La Carte, mode d'emploi.* Paris: Fayard, 1987.

Cage, John. *Score without Parts (40 Drawings by Thoreau)/ Twelve Haiku.* Oakland: Crown Point Press, 1978.

Careri, Francesco. *Walkscapes: La marche comme pratique esthétique.* Arles: Actes Sud, 2013.

Lewis Carroll. *The Hunting of the Snark.* London: Macmillan, 1876.

Certeau, Michel de. *L'Invention du quotidien, vol. 1: Arts de faire.* New ed. Paris: Gallimard, 1980.

Clerc, Thomas. *Paris, Musée du XXI*[e] *siècle: Le dixième arrondissement.* Paris: Gallimard, 2007.

Charbonnier, Pierre, Bruno Latour, and Baptiste Morizot. "Redécouvrir la Terre." *Tracés, revue de sciences humaines*, no. 33 (2017).

Coccia, Emanuele. *The Life of Plants: A Metaphysics of Mixture.* Trans. Dylan J. Montanari. Cambridge: Polity, 2017.

Cosgrove, Denis, ed. *Mappings.* London: Reaktion Books, 1999.

Cosgrove, Denis. "Historical Perspectives on Representing and Transferring Spatial Knowledge." In *Mapping in an Age of Digital Media: the Yale Symposium, ed.* Mike Silver and Diana Balmori, 128–137. New York, Wiley-Academy, 2003.

Crampton, Jeremy. "Cartography: Performative, Participatory, Political." *Progress in Human Geography* 33, no. 6 (May 21, 2009): 840–848.

Crang, Mike, and Nigel Thrift, eds. *Thinking Space: Critical Geographies.* London: Routledge, 2000.

Crary, Jonathan. *Techniques of the Observer: On Vision and Modernity in the Nineteenth Century.* Cambridge, MA: MIT Press, 1990.

Crary, Jonathan. *24/7: Late Capitalism and the Ends of Sleep.* London: Verso, 2014.

Damisch, Hubert. "La grille comme volonté et comme représentation." In *Cartes et figures de la Terre*, 30–40. Paris: Centre Georges Pompidou/CCI, 1980.

Danowski, Deborah, and Eduardo Viveiros de Castro. *The Ends of the World.* Trans. Rodrigo Nunes. Cambridge: Polity, 2017).

Daston, Lorraine, and Peter Galison. *Objectivity.* Chicago: University of Chicago Press, 2007.

Davis, Mike. *Dead Cities.* New York: The New Press, 2002.

Debaise, Didier. *L'Appât des possibles.* Dijon: Presses du réel, 2015.

Deleuz, Gilles, and Félix Guattari. *A Thousand Plateaus: Capitalism and Schizophrenia.* Trans. Brian Massumi. Minneapolis: University of Minnesota Press, 1987.

Dupont, Daniel. "Gilles Deleuze sur Leibniz—Le point de vue (1986)." YouTube video, January 28, 2017. https://www.youtube.com/watch?v=gVpcNjE3vY8.

Descola, Philippe. *Par-delà nature et culture.* Paris: Gallimard, 2005.

Descola, Philippe. "Figures des relations entre humains et non-humains" and "Ontologie des images." Class lecture, Anthropologie de la Nature, Collège de France, Paris, 2000–2017.

Despret, Vinciane. "En finir avec l'innocence. Dialogue avec Isabelle Stengers et Donna Haraway." In *Penser avec Donna Haraway, ed.* Elsa Dorlin and Eva Rodriguez. Paris: PUF, 2012.

Dörk, Marian, Rob Comber, and Martyn Dade-Robertson. "Monadic exploration: Seeing the whole through its parts." In *CHI 2014: Proceedings of the SIGCHI Conference on Human Factors in Computing Systems*, 1535–1544. New York: ACM, 2014.

Dodge, Martin, Rob Kitchin, and Chris Perkins, eds. *Rethinking Maps: New Frontiers in Cartographic Theory.* London: Routledge, 2009.

Dufour, Diane and Jean-Yves Jouannais, eds. *Topographies de la guerre.* Göttingen: Steidl/Paris: Le Bal, 2011.

Farinelli, Franco. *De la raison cartographique. Trans.* Katia Bienvenu. Paris: CTHS, 2009.

Fuller, R. Buckminster. *Operating Manual for Spaceship Earth.* Zurich: Lars Müller, 2008.

Gaz, Stan. *Sites of Impact: Meteorite Craters Around the World.* Princeton: Princeton Architectural Press, 2009.

Glon, Éric. "Accéder à la carte: L'exemple des cartographies autochtones chez les Lil'wat."

In *Opérations cartographiques, ed.* Jean-Marc Besse and Gilles A. Tiberghien, 218–229. Arles: Actes Sud/ Versailles: ENSP, 2017.

Grégoire, Axelle. "Cartographier avec le vivant, une redécouverte de la plasticité des territoires." *Urbia, les cahiers du développement urbain durable*, no. 22 (March 2019): 127–145.

Hache, Émilie, ed. *De l'univers clos au monde infini.* Bellevaux: Éditions Dehors, 2014.

Hache, Émilie. *Reclaim: Recueil de textes écoféministes.* Paris, Cambourakis: 2016.

Haraway, Donna. *SF: Speculative Fabulation and String Figures.* Ostfidern: Hatje Cantz, 2012.

Haraway, Donna. "Situated Knowledges: The Science Question in Feminism and the Privilege of Partial Perspective." *Feminist Studies* 14, no. 3 (1988): 575–599.

Haraway, Donna. *Staying with the Trouble: Making Kin in the Chthulucene.* Durham: Duke University Press, 2016.

Harmon, Katharine. *The Map as Art: Contemporary Artists Explore Cartography.* New York: Princeton Architectural Press, 2009.

Hofmann, Catherine, ed. *Artistes de la carte, de la Renaissance au XXIe siècle: L'explorateur, le stratège, le géographe.* Paris: Autrement, 2012.

Hofmann, Catherine, ed. *L'Âge d'or des cartes marines: Quand l'Europe découvrait le monde.* Paris: Seuil/BNF, 2014.

Hooke, Robert. *Micrographia.* London: Royal Society, 1665.

Humboldt, Alexander von. *Cosmos: A Sketch of a Physical Description of the Universe.* Ed. Edward Sabine. Cambridge: Cambridge University Press, 2011.

Ingold, Tim. *A Brief History of Lines.* London: Routledge, 2016.

Ishigami, Junya. *Another Scale of Architecture.* Kyoto: Seigensha, 2010.

Jacob, Christian. *L'Empire des carte:, Approche théorique de la cartographie à travers l'histoire.* Paris: Albin Michel, 1992.

Jullien, François. *Vivre de paysage, ou L'impensé de la raison.* Paris: Gallimard, 2014.

Kopenawa, Davi, and Bruce Albert. *The Falling Sky: Words of a Yanomami Shaman. Trans.* Nicholas Elliott and Alison Dundy. Cambridge, MA: Harvard University Press, 2013.

Küster, Ulf, ed. *Theatrum Mundi—Die Welt als Bühne.* Munich: Minerva, 2003.

Lacoste, Yves. *La Légende de la Terre.* Paris: Flammarion, 1996.

Latour, Bruno, Valérie November, and Eduardo Camacho-Hübner. "Entering a Risky Territory: Space in the Age of Digital Navigation." *Environment and Planning* 28, no. 4 (2010): 581–599.

Latour, Bruno. *Down to Earth: Politics in the New Climatic Regime.* Cambridge: Polity, 2018.

Latour, Bruno. *Facing Gaia: Eight Lectures on the New Climatic Regime.* Cambridge: Polity, 2017.

Latour, Bruno. "Some Advantages of the Notion of 'Critical Zone' for Geopolitics." Lecture, Geochemistry of the Earth's Surface Meeting, Paris, August 18–23, 2014.

Lenton, Tim. *Earth System Science: A Very Short Introduction.* Oxford: Oxford University Press, 2016.

Lévy, Jacques, ed. *A Cartographic Turn: Mapping and the Spatial Challenge in Social Sciences.* Lausanne: EPFL Press, 2016.

Lovelock, James. *Gaia: A New Look at Life on Earth.* 3rd ed. Oxford: Oxford University Press, 2000.

Lussault, Michel. *De la lutte des classes à la lutte des places.* Paris: Grasset, 2009.

Macchi, Giulio, ed. *Cartes et Figures de la Terre.* Paris: Centre Pompidou / CCI, 1980.

Margulis, Lynn, and Dorion Sagan. *Microcosmos: Four Billion Years of Microbial Evolution.* Berkeley: University of California Press, 1986.

Misrach, Richard, and Kate Orff. *Petrochemical America.* New York: Aperture Press, 2012.

Monsaingeon, Guillaume. *Mappamundi: Art et cartographie.* (arseille: Parenthèses, 2013.

Morizot, Baptiste. *Les Diplomates: Cohabiter avec les loups sur une nouvelle carte du vivant.* Marseille: Wildproject Éditions, 2016.

Morizot, Baptiste. "Nouvelles alliances avec la Terre: Une cohabitation diplomatique avec le vivant." *Tracés: Revue de sciences humaines*, no. 33 (2017).

Morizot, Baptiste. *Sur la piste animale.* Arles: Actes Sud, 2018.

Myers, Natasha. "Photosynthetic Mattering: Rooting into the Planthroposcene." In *Moving Plants*, ed. Line Marie Thorsen. Næstved: Narayana Press, 2017.

Mockers, Elizabeth. *Géographes arabo-musulmans du Xe au XIVe siècle: La côte méditerranéenne de la péninsule Ibérique chez les géographes arabo-musulmans et les cartographes chrétiens du Xe au XIVe siècle.* Colomars: Melis, 2006.

Palsky, Gilles. "Carte, temps et récit." In *Opérations cartographiques*, ed. Jean-Marc Besse and Gilles A. Tiberghien, 57–69. Arles: Actes Sud/Versailles: ENSP, 2017.

Palsky, Gilles. *Des chiffres et des cartes: Naissance et développement de la cartographie quantitative française au XIXe siècle.* Paris: CTHS géographie, 1996.

Pelletier, Monique, ed. *Couleurs de la Terre: Des mappemondes médiévales aux images satellitales.* Paris: Seuil/Bibliothèque nationale de France, 1998.

Pelletier, Monique, ed. *Géographie du monde au Moyen Âge et à la Renaissance: Actes de la 12e Conférence internationale d'histoire de la cartographie.* Paris: CTHS, 1989.

Pickles, John. *A History of Spaces: Cartographic Reason, Mapping, and the Geo-Coded World.* London: Routledge, 2004.

Piron, Sylvain. *Dialectique du monstre: Enquête sur Opicino de Canistris.* Brussels: Zones sensibles, 2015.

Reclus, Élisée, Alexandre Chollier, and Federico Ferretti, eds. *Écrits cartographiques.* Geneva: Éditions Héros-Limite, 2016.

Rifkin, Jeremy. *The Third Industrial Revolution: How Lateral Power Is Transforming Energy, the Economy, and the World.* New York: Palgrave Macmillan, 2011.
Rocher, Yann, ed. *Globes: Architecture et sciences explorent le monde.* Paris: Norma, 2017.
Rockström, Johan, Will Steffen, Kevin Noone, Åsa Persson, F. Stuart Chapin III, Eric F. Lambin, et al. "A Safe Operating Space for Humanity." *Nature* 461 (September 2009): 472–475.
Rollot, Mathias. *L'Obsolescence: Ouvrir l'impossible.* Geneva: MétisPresses, 2016.
Rosa, Hartmut. *Accélération: Une critique sociale du temps.* Paris: La Découverte, 2010.
Rossi, Aldo. *The Architecture of the City. Trans.* Diane Ghirard and Joan Ockman. Cambridge, MA: MIT Press, 1982.
Stengers, Isabelle. *In Catastrophic Times: Resisting the Coming Barbarism. Trans.* Andrew Goffey. London: Open Humanities Press, 2015.
Stengers, Isabelle. "Penser à partir du ravage écologique." In *De l'univers clos au monde infini*, ed. Émilie Hache. Bellevaux: Éditions Dehors, 2014.
Schüler, Chris, ed. *Dessiner le monde: Atlas de la cartographie du XVIe à 1914.* Paris: Place des Victoires, 2010.
Thorsen, Line Marie, and Anette Vands. "Can We Land on Earth? An Interview with Bruno Latour." In *Moving Plants*, ed. Line Marie Thorsen. Næstved: Narayana Press, 2017.
Tiberghien, Gilles A. *Finis terrae: Imaginaires et imaginations cartographiques.* Paris: Bayard, 2007.
Toledo, Camille de, Aliocha Imhoff, and Kantuta Quirós. *Les Potentiels du temps: Art et politique.* Paris: Manuella éditions, 2016.
Tresch, John. "Cosmogram." In *Cosmogram*, ed. Jean-Christophe Royoux and Melik Ohanian, 67–76. New York: Sternberg, 2005.
Tsing, Anna Lowenhaupt *The Mushroom at the End of the World: On the Possibility of Life in Capitalist Ruins.* Princeton: Princeton University Press, 2015.
Tsing, Anna Lowenhaupt, Heather Swanson, Elaine Gan, and Nils Bubandt, eds. *Arts of Living on a Damaged Planet: Ghosts and Monsters of the Anthropocene.* Minneapolis: University of Minnesota Press, 2017.
Uexküll, Jakob Johann von. *A Foray into the Worlds of Animals and Humans: With A Theory of Meaning. Trans.* Joseph D. O'Neil. Minneapolis: University of Minnesota Press, 2013.
Usher, Phillip John. *Exterranean: Extraction in the Humanist Anthropocene.* New York: Fordham University Press, 2019.
Vasset, Philippe. *La Conjuration.* Paris: Fayard, 2013.
Vasset, Philippe. *Un livre blanc: Récit avec cartes.* Paris: Fayard, 2007.
Vidalou, Jean-Baptiste. *Être forêts: Habiter des territoires en lutte.* Paris: Zones/La Découverte, 2017.
Weller, Richard, Claire Hoch, and Chieh Huang. Atlas for the End of the World. http://atlas-for-the-end-of-the-world.com. Accessed March 5, 2021.
Wohlleben, Peter. *La Vie secrète des arbres: Ce qu'ils ressentent, comment ils communiquent, un monde inconnu s'ouvre à nous.* Paris: Les Arènes, 2017.
Zalasiewicz, Jan. "The Extraordinary Strata of the Anthropocene." In *Environmental Humanities: Voices from the Anthropocene*, ed. Serpil Oppermann and Serenella Iovino. Lanham, MD: Rowman & Littlefield, 2016.
Zalasiewicz, Jan, Will Steffen, Reinhold Leinfelder, Mark Williams, and Colin Waters. "Petrifying Earth Process: The Stratigraphic Imprint of Key Earth System Parameters in the Anthropocene." *Theory, Culture & Society* 34, nos. 2–3 (2017): 83–104.

AUTHOR BIOS

Frédérique Aït-Touati is a historian of science and a theater director. Trained in literary studies at the École normale supérieure and in the history of science at Cambridge University, she explores the links between science, literature, and politics and is particularly interested in the heuristic and cognitive potential of storytelling and fiction. She taught literature at Oxford University from 2007 to 2014, before returning to France as a researcher at the French National Center for Scientific Research. Her published works include *Fictions of the Cosmos: Science and Literature in the Seventeenth Century* (2011) and *Le Monde en images* (2015). She has worked for many years with the philosopher Bruno Latour; together, they have developed various forms of theatrical and performative writing, including the show *Gaïa Global Circus* (2013–2016), the simulation *Théâtre des Négociations/Make It Work*, the show *Inside*, and the performance *Back to Earth* (Centre Pompidou, 2019). She also teaches at the School for Advanced Studies in the Social Sciences (EHESS) and directs the Program for Experimentation in Political Arts (SPEAP).

Alexandra Arènes is an architect at SOC (Société d'Objets Cartographiques, s-o-c.fr). Since 2018, she has been a doctoral researcher in the Manchester Architecture Research Group at The University of Manchester, working on a project titled "Architectural Design at the Time of Anthropocene: A Gaia-graphic Approach to the Critical Zones" with Professor Albena Yaneva. Her research interests include the impact of the Anthropocene on landscape studies and the collaboration between the arts and sciences to renew methods for envisioning landscape.

Axelle Grégoire is an architect who has worked for several years as a project manager in the urban planning department of the landscape design agency BASE in Paris. She is trained in various traditional skills (etching, woodworking), as well as digital tools. She develops projects that contribute to the renewal of the representation of territories and their rewriting, within *S.O.C.* (Société d'Objet Cartographique) and its studio Omanoeuvres. In the pursuit of a project on the "Forest-City" (recipient of the third Villa Le Nôtre research and landscape design residency), since 2020 she has been a doctoral student at the CESCO (TEEN group) of the Muséum National d'Histoire Naturelle (MNHM) under the direction of Anne-Caroline Prévot. She also teaches at the Art and Design Superior School of Valenciennes (ESAD).

Studio SOC was cofounded by Alexandra Arènes, Axelle Grégoire, and Soheil Hajmirbaba in 2016. The studio designed the installation "CZO space" at the ZKM Museum for Art and Media in Karlsruhe, Germany, for the 2020 exhibition *Critical Zones: Observatories for Earthly Politics*, curated by Bruno Latour. The installation is the result of a close collaboration between artists and the French scientific network of Critical Zones (OZCAR). The studio also contributes to scenography with Frédérique Aït-Touati (*INSIDE, Back to Earth*) and develops political workshops for the *Where to Land?* project with Bruno Latour.

The *Terra Forma* team continues to develop the *Terra Forma* project with cartographic workshops in design, art, and architecture schools around the world, and through a new research project on cosmopolitical mappings.

ACKNOWLEDGMENTS

This book owes much to discussions with our colleagues and friends. We would particularly like to thank Emanuele Coccia, Bruno Latour, Sonia Levy, Baptiste Morizot, Yann Rocher, and Simon Schaffer, whose comments, generous suggestions, and inspiring proposals have nurtured this book. We would also like to thank David Bornstein, Søheil Hajmirbaba, Vincent Imfeld, and Luc Hélénon for their careful proofreading. Without the high standards and the kindness of our B42 editors, Alexandre Dimos, Julia Lamotte, and Tony Côme, this book would not have been possible.

One inspiration for this project is the work carried out within the Baseland agency. We would like to thank our various collaborators, landscapers, urban planners, and architects, as well as the contracting authorities with whom we have worked on the development projects referenced in these pages.

This book was written in part during a residency at the Petite Égypte bookshop in Paris, hosted by Alexis Argyroglo as part of a writing grant from the Île-de-France region. We also received support from the Initiative for Strategic and Innovative Research project Origins and Conditions for the Emergence of Life (IRIS OCAV) at the PSL observatory in Paris. Finally, our thanks go to Jérôme Gaillardet and his "critical zonist" colleagues, researchers investigating the Critical Zone whose journeys we have sometimes followed.

Unless otherwise stated, all illustrations were created by the authors.
Source of the maps on pages 52, 76, and 100: geoportail.gouv.fr

This book was designed in Paris by deValence.
Printed and bound by Musumeci, Italy.

10 9 8 7 6 5 4 3 2 1

Library of Congress Cataloging-in-Publication Data

Names: Aït-Touati, Frédérique, 1977– author. | Arènes, Alexandra, author. | Grégoire, Axelle, author. | Latour, Bruno, author of foreword.
Title: Terra forma: a book of speculative maps / Frédérique Aït-Touati, Alexandra Arènes, and Axelle Grégoire ; foreword by Bruno Latour; translated by Amanda DeMarco.
Other titles: Terra forma. English
Description: Cambridge, Massachusetts ; London, England: The MIT Press, [2022] | "This book was first published in 2019 by Éditions B42, Paris. © B42, Frédérique Aït-Touati, Alexandra Arènes, and Axelle Grégoire"—T.p. verso.

Identifiers: LCCN 2021013274 |
ISBN 9780262046695 (Hardcover)
Subjects: LCSH: Cartography. | Earth (Planet)—Maps.
Classification: LCC GA101.5 .A3813 2022 | DDC 912—dc23
LC record available at https://lccn.loc.gov/2021013274

The writing of this book was supported by PSL within the framework of IRIS OCAV for the Critical Zone research project (ANR-IDEX-0001-02). It also benefited from the support of the Center for Research on Arts and Language at the School for Advanced Studies in the Social Sciences, the Baseland agency, the shaā studio of architecture, SOC: Society of Cartographic Objects, and the Centre national du livre.

This translation was supported by the Centre national du livre, and the CRAL - École des hautes études en sciences sociales.

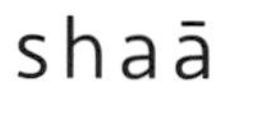